CREATIVITY AND INTUITION

CREATIVITY
AND
INTUITION

A Physicist Looks at East and West

by
Hideki Yukawa

translated by
John Bester

KODANSHA INTERNATIONAL LTD.
Tokyo, New York & San Francisco

Section 3, chapter 1, "Intuition and Abstraction in Scientific Thinking," was first published in *Frontiers of Modern Scientific Philosophy and Humanism: The Athens Meeting, 1964*
© Elsevier, Amsterdam, 1966

Section 3, chapter 2, "Creative Thinking in Science," was first published in *Man and His World/Terre des Hommes: The Noranda Lectures at Expo 67*
© University of Toronto Press, 1968

Section 4, chapter 2, "Meson Theory in Its Developments"
© The Nobel Foundation, 1964

Distributors:

UNITED STATES: *Harper & Row, Publishers, Inc.*
10 East 53rd Street, New York, New York 10022

CANADA: *Fitzhenry & Whiteside Limited*
150 Lesmill Road, Don Mills, Ontario

GENERAL AND SOUTH AMERICA: *Feffer & Simons Inc.*
31 Union Square, New York, New York 10003

EUROPE: *Boxerbooks Inc.*
Limmatstrasse 111, 8031 Zurich

THAILAND: *Central Department Store Ltd.*
306 Silom Road, Bangkok

HONG KONG: *Books for Asia Ltd.*
379 Prince Edward Road, Kowloon

THE FAR EAST: *Japan Publication Trading Company*
P.O. Box 5030, Tokyo International, Tokyo

Published by Kodansha International Ltd., 2-12-21, Otowa, Bunkyo-ku, Tokyo 112, Japan and Kodansha International/USA, Ltd., 10 East 53rd Street, New York, New York 10022 and 44 Montgomery Street, San Francisco, California 94104. Copyright 1973, by Kodansha International Ltd.
LCC 78-174222
ISBN 0-87011-177-9
JBC 1040-783462-2361

First edition, 1973

CONTENTS

TRANSLATOR: *With the exception of the chapters in sections 3 and 4 on "Intuition and Abstraction in Scientific Thinking," "Creative Thinking in Science," "Meson Theory in Its Developments," and "Space-Time and Elementary Particles," this book was translated from the Japanese by John Bester.*

INTRODUCTION

In facing nature, all human beings are equal. We may not understand each other's different languages or different cultures and philosophies, we may not follow the same ways of living, but scientists all over the world understand the scientific language, they come to the same scientific conclusions, they appreciate, enjoy, and value scientific insights with the same interest and enthusiasm.

This book deals with the miracle of science, with the miracle of human thinking that is capable of penetrating into the inner workings of nature. Einstein said once that the most incomprehensible fact is the fact that nature is comprehensible. Part of that miracle is the fact that it is equally comprehensible for any scientifically trained human being, whatever his or her origin or cultural background.

The Japanese cultural tradition is far removed from the Western way of thinking. Its history and philosophy have followed very different paths. It is of special interest, in particular for the Western reader, to learn how an eminent Japanese scientist is thinking about science, how he approaches the problems, how his scientific life is rooted in the philosophies of his culture, and what were the motivations and urges that brought him to science. In following his ideas and his tales, we learn much about Oriental thinking and feeling; we also learn about the Western character and we learn most about what is common to all of us. It is always profitable to look at old problems from a different point of view. Things become much clearer and we are more apt to distinguish the essential from the nonessential.

In the first chapter, Yukawa tells us about his education and how he became a theoretical physicist. He grew up in a milieu different from a Western environment, but in the making of a theorist most elements are similar all over the world. He traces his choice of subject to his aversion to the world of social and personal contacts, an aversion which brought him to the apparently safer but not less real world of mathematical description of nature. He found out, later in his life, as did many of his colleagues, that even the most abstract theories are linked in one way or another to human society. He, more than most of his colleagues, recognized the importance of these connections and the dangers inherent in thoughtless applications of scientific results, and he used his influence and reputation for the maintenance of world peace and against the use of science for war and suppression.

His Japanese-Chinese traditions, however, made him especially aware of the contrast between logical rigor and the role of intuition and imagination in science and in other human endeavors. It is the problem of logical deduction versus intuitive thinking which always appears in his deliberations. Chinese thinking emphasizes the intuitive component—the importance of inspiration in scientific thinking. The fundamental desire for harmony in the world appears as one of the mainsprings of natural philosophy. He tells us about the ideas of those Chinese philosophers who had the greatest influence on him: Laotse, Chuangtse, and Motse. A special flavor emanates from these chapters as the reader finds himself immersed in a world of thoughts, strange and familiar at the same time, dealing with the deepest questions of mankind in ways which are both simple and sophisticated. We hear Laotse and Chuangtse say that man divorced from nature could not be happy and man's power to resist nature is hopelessly puny. This attitude is different from the tenets of scientific civilization, which pretends to have conquered our impotence in the face of nature in the raw. But Yukawa reminds us of recent developments when he quotes Laotse saying that "Heaven and Earth are without compassion; they see all things as straw dogs."

Then there is this fable about the man who wanted to get rid of his shadow and of his footprints by running ever faster. And Chuangtse remarked: "Foolish man: if he had stayed in the shade, he would have had no shadow; if he had been still, there would have been no footprints." Is this spirit opposed to science? It is, but in a deeper way it is not, since science also is a form of contemplative wisdom.

Some of the thoughts remind one of Niels Bohr's philosophy of complementarity. The most striking example is Laotse's statement: "The kind of way that can be expressed clearly is not the absolute and immutable way." Bohr has said that there is a complementarity between clarity and truth (*Klarheit* and *Wahrheit*); any attempt to express a thought in words involves some change, some irrevocable interference with the essential idea.

How wonderful is the description of two complementary attitudes towards science in the fable about "the happy fish"! It concerns the outlook of Huitse, who will not accept anything that is not exactly proven, and the outlook of Chuangtse, who cherishes intuitive and vaguely conceived ideas. Yukawa asks his colleagues to which of the two they are related in their thoughts. There are too many Huitses in modern physics!

The thoughts of Motse deal with the problems of ethics and

religion. Again it is at the same time strange and familiar to learn about *chien-ai*, a doctrine that teaches one to care for the fate of one's fellow man but condemns the idea of self-sacrifice for others. Humanity is served best if everyone thinks of his own benefits as well as of those of others. "The man who benefits others will almost certainly reap benefits from them. If one hates others one will be hated in return." And Motse unfortunately is right when he says that "those responsible for government do not make it [*chien-ai*] the basis of their rule."

The second part of the book deals with the question of creative thinking in science. It contains many important and illuminating thoughts on this subject which is central to Yukawa's way of thinking and searching. We find this wonderful description of creativity: "One has some place that is dark, or obscure, or vague, or puzzling within oneself, and one tries to find some light in it. Then, when one has found a ray of light, one tries to enlarge it little by little so that the darkness is gradually dispelled." Yukawa searches for the roots of creativity and is deeply concerned with the present state of modern physics. There are several trends that worry him: "There seems to exist a general feeling of estrangement of science from other cultural activities such as philosophy and literature." He ascribes this split in our cultures to the secondary role to which intuition and imagination have been relegated. "Most of the physicists of today are at a loss to know what to do with the richness and complexity of the newly explored world of subnuclear physics. It seems as if present-day physicists have lost the gift of foresight inherited from their forerunners." He sees the causes of this change of attitudes in the fact that "to some physicists of the younger generation theoretical physics is reduced to mathematics of complex functions of complex variables supplemented by mathematics of abstract groups . . . this one-sided trend to abstraction lacks something which is very important to creative thinking . . . abstraction cannot work by itself, but has to be accompanied by intuition or imagination." He complains that "during the sixty years of this century, theoretical physics has become less and less romantic . . . we are now in an anti-romantic era." He makes the observation that today "one of the prime aims of fundamental physics is to obtain a great number of data from a big accelerator and then to put them in a high-speed computer to analyze and compare the results with theoretical formulae."

It is good to see thoughts like these expressed in such strong terms. We may, however, face here a problem of the famous generation gap. Perhaps the scientists of the younger generation also have warm and emotional feelings towards their subject, but they express them in different terms. We must not forget that the older generation of physicists was to some extent spoiled by the great breakthrough of theoretical physics at the beginning of our century. Such great leaps forward do not happen all the time; they may occur once a century only. Yukawa mentions the strange phenomenon that "geniuses appear in batches," as has happened in Newton's time, in Faraday's and Maxwell's time, and now in Bohr's, Einstein's, and Yukawa's time. There are periods in which the accumulated knowledge is ripe for new insights; in such moments the situation may have created geniuses rather than the other way round. One should also keep in mind that there are periods in the development of science when ingenuity and imagination express themselves less in theoretical syntheses and more in terms of new experimental ideas and possibilities. These periods alternate with those of new theoretical insights; certainly today we are in one of these periods and we witness tremendous achievements of that other kind of ingenuity which has led to the discoveries of new and unexpected phenomena at high energies, at low temperatures, and at large distances.

The essays on theoretical physics illustrate directly the supra-national aspect of physics. Here we listen to Yukawa the physicist, expounding his views and ideas about modern particle physics. His thoughts often are novel and different from the beaten path, pointing towards new ways of solving the riddles which the art of experimentation has revealed to us. Here Yukawa speaks as a great physicist and no longer as a representative of the Chinese-Japanese culture.

His speech delivered at the occasion of his Nobel award reminds us of the greatness of the discovery for which he received this distinction. We are impressed by the simplicity of approach and by the unfailing intuition of a great master. Only a few facts were known at that time regarding the properties of nuclear forces. He considered the observed short-range of these forces, saw the fundamental significance of this observation, and made his prophetic prediction of the existence of a new kind of particle: the meson. On the basis of a few observations, he anticipated the new world of mesonic phenomena which was indeed discovered bit by bit in the decades following his inspiration.

The last section of his essays is entitled "On Peace." A great physicist and a great human being is deeply troubled by the predicament of mankind endowed with too much knowledge and power. We recall that an aversion to the intractability of the problems of human relations brought the young Yukawa towards the impersonal problems of nature and its laws. The mature Yukawa recognized that science is part of human relations and that modern scientific work involves him and other scientists with society to an even greater extent than other people. Only five years after his insight into the nature of the nuclear force, the same force was used to unleash huge amounts of energy for the purpose of destruction of human life and habitation. The twists and turns of history forced most of his colleagues all over the world to use their ingenuity not for the understanding of nature but for turning dramatic new discoveries into deadly weapons of war. Japan was the first and only country by fate and historic involvement that has been the victim of such a cruel perversion of nuclear science. Of necessity, therefore, Yukawa was among the most outspoken scientists in the postwar era, who used their influence and reputation against the misuse of science in war and for a world of peace among nations. The articles and essays in the last section of this book bear witness to the depth of his feelings and to the persuasiveness of his words. But they also demonstrate how difficult is the task of turning the age-old tide of human ambition and aggression. More than political measures, treaties, and professions of good will are necessary; a new way of thinking about human values is needed. Yukawa reports his impressions at one of the Pugwash meetings, in which scientists from all over the world assemble to discuss measures against war and for a better use of science. He sympathizes with the efforts and is ready to help, but he fears that the aims were not set high enough. In reading this personal account of his feelings when attending the meetings, one is deeply struck by his concern about the large gap between the superhuman task and the frailties of human endeavors, even if they are done with the best of intentions. He quotes the words of Max Born: "Clever, rational ways of thinking are not enough. The danger of mass slaughter . . . can only be overcome by moral conviction, by a determination to replace national prides and prejudices with human love." But Yukawa has not given up hope. In his concluding essay he expresses his trust in the power of rational thinking. He believes that "we have the possibility . . . of gradually bringing a wider area of our humanity within the scope of rational consideration." And he sums up his remarks with these wonderful words of a man who is deeply involved in action, contemplation, and hope: "Resignation may be important for human beings; the time may come when I have to give up. But I have not given up yet."

VICTOR F. WEISSKOPF

PREFACE

Born a Japanese, I have spent the whole of my sixty-six years, with the exception of five years' stay in America, in Japan. Yet partly because of the fact that I became a physicist, I have in practice developed an outlook that is far more cosmopolitan than that of the average Japanese. This cosmopolitan quality, having been acquired chiefly through the medium of modern science, which is universal in its application, relates chiefly to the nations of the West.

However, seen in a broader historical perspective, the Japanese as a whole had already acquired an international quality, albeit of a rather different kind, at a very early date, a quality that derived mainly from their contact with the culture of China. At a certain stage in the ancient period, the culture of Korea, both in itself and as an indirect reflection of Chinese culture, also had a great significance. Another factor, of course, that enhanced this cosmopolitan quality was the entry into Japan, and the popularization there, of the Buddhism of India, which came via China and Korea. Nevertheless, because most of these influences from the cultures of other Oriental nations, having begun in ancient times, and extended over a very long period of years, have become completely assimilated in Japan, the modern Japanese has ceased to feel any strong cosmopolitan element in them.

Where the acquisition of cosmopolitan qualities via the culture and ideas of the East is concerned, I have, again, found myself in slightly different circumstances from the majority of the Japanese. As I frequently mention in parts 1 and 2 of this volume, my acquaintance with the Chinese classics had begun even before I went to primary school. This would have been the normal experience of any son of the intellegentsia during the Edo and early Meiji periods, but it was a type of education that had almost ceased to exist by the time I was born.

It is not easy to decide just what effect such an experience had on me later as I took up the study of physics, that most typical of all modern sciences. What is certain, however, is that it had an important influence on my outlook on life and on the universe. The influence of the ideas of Laotse and Chuangtse was particularly great; this was unusual in Japan, since unlike Buddhist ideas, which over long centuries had come to permeate the outlook of the Japanese as a whole, or the Confucianism—more specifically, the Neo-Confucianism—that became the official learning of the Edo period, Taoism had aroused little interest in Japan at any stage in its history. Historically a kind of antithesis to Confucianism, it comprised at the same time ideas that are in conflict with those of modern science. It followed that, for a long while after I took up physics, those ideas remained beneath the surface of my consciousness, though this does not necessarily mean that they were not at work somewhere in my subconsciousness.

Leaving aside any conclusions that might be drawn from a more detailed investigation of this point, it cannot be denied that with the increasing doubts about scientific civilization that have developed since the appearance of the atomic bomb and the realization that scientific civilization can no longer be considered omnipotent, the ancient philosophers of the East have come to acquire a new significance. Many readers may feel that the essays on peace in part 5 tend to be overidealistic, but for the scientist who knows that science is not almighty the views expressed may, on the contrary, seem almost trite.

So far, I have stressed the cosmopolitan aspect of my upbringing, but it goes without saying that as a Japanese I have a deep interest in the culture and ideas that are indigenous to Japan, and have unconsciously been subjected to an overwhelming influence from that quarter. In the various fields of art and literature in particular, Japan has, over its long history, achieved a great deal that is of international rank, and I have been an avid reader of the Japanese classics ever since I was a boy. In more recent years, I have myself written many essays on the subject. In the present volume, however, I decided to include only one of them, a reflection on Japan's most celebrated classical novel, *The Tale of Genji*. The point I set out to make in this essay was that the novel constitutes—and this is very much related to the fact that the author was a woman—a world where light and shadow stand in reverse positions from in the world of physics, and that human beings have the ability to dwell in either of these worlds.

I return again to the question of cosmopolitanism, since from the time I became a physicist my interest naturally turned also to the ancient Greek thinkers who first gave birth to physics. In various ways I felt a particular affinity with Epicurus, and I wrote a short essay principally about him, which I have included with the essay on *Genji* in part 2. In 1964—shortly after writing this essay, though there was no direct connection with it—I was given an unexpected chance to visit Greece, and I have included in part 3 the lecture on "Intuition and Abstraction in Scientific Thinking" that I gave in Athens on that occasion. The next article, "Creative Thinking in Science," formed part of the

Noranda Lectures, a series delivered in connection with the World Exposition held in Montreal in 1967. The two last-named lectures, which were written originally in English, examine the question of how creativity makes itself apparent in the development of science (the following piece, "The Conception and Experience of Creativity," is a somewhat more detailed discussion of the same question).

The articles in part 4 deal with modern physics as such and the modern physicist's apprehension of nature. One of them, "Meson Theory in Its Developments," is a lecture given in Stockholm, and was written in English. The essay that follows, "Space-Time and Elementary Particles," was originally written in Japanese and then translated with the expectation that it would be published in a book to be a collection of essays by scientists; however the plans for that book fell through and it is published here for the first time in English. Apart from these four articles, all the pieces in this book were translated by Mr. John Bester from Japanese written or spoken for the Japanese reader or for a Japanese audience. Not only are the structures of the Japanese and English languages extremely different to begin with, but the matters written or spoken inevitably include some that are difficult to understand or of little interest for the non-Japanese. "On Learning and Life" in particular, being the record of a dialogue between myself and an old acquaintance, Professor Takeshi Inoue, must have been particularly difficult to turn into English, both stylistically and in its content. That a coherent work in English has emerged in spite of these difficulties is thanks to the labors of Mr. Bester, to whom I extend my sincere thanks.

When the English translation was more or less ready, I sent it to an old friend, Professor Victor F. Weisskopf, with a request for a foreword. His response was enthusiastic; I am very happy that despite the great differences of cultural background he should have shown such a deep understanding of the contents of this book, and I express my sincere gratitude to him.

The first plans for this book were mooted by Kodansha International a considerable while ago; there was practically no precedent in Japan for this kind of work, so that President Toshiyuki Hattori, Managing Director Saburō Nobuki, and the editorial staff have been put to considerable trouble in the course of preparing it for publication; Mr. Gyō Furuta in particular devoted a great deal of energy to the early stages of planning the book. I wish here to express my gratitude to them all.

HIDEKI YUKAWA

I ON LEARNING
AND LIFE [1968]

Character and Course

Everyday Life and the Road to Research

INOUE[1]: For some time now, Dr. Yukawa, you have been writing various pieces of a more or less autobiographical nature, and recently you've been laying a good deal of stress on human creativity. Your approach to the problem shows a strong tendency to treat life and scholarship as things at once distinct from each other and inseparable; in this respect, I feel it will not only, in one sense, serve to stimulate young people's interest in scholarship but is also extremely instructive in itself. So perhaps you could say something about your approach to scholarship with special reference to the problems of human life.

YUKAWA: I like to look at it in this way. If you ask what view of life or the universe lies behind my recent preoccupation with the question of creativity, I feel that although man has a sense of purpose, he also has the awareness that his surroundings—the world in which he himself is supposed to be acting with that sense of purpose—always contain some element of the unknown, and that the unknown part is extremely important. It means that when man himself does research, or discovers something, or confirms something, the world itself changes for him at the same time. Or it may be that the surroundings in which he finds himself are changed, not by himself, but by something done by others, or by natural phenomena. Thus man is always in a changing situation that he cannot foresee completely. Nevertheless, to live within such a situation is not merely aimless. The aims include some that cannot be dismissed as merely temporary. There is something that does not, as it were, go on changing remorselessly. The aims, in other words, include in themselves something that is permanent, that does not change throughout one's life. That's how I see things. To look back over my own life, back to the time before I became a scholar, I first set myself an aim in life—to study, and in particular to study physics—when I was in my final grades at high school.

INOUE: Yes—though I imagine you had also been influenced by your surroundings since the time you were a child.

YUKAWA: If one goes back still earlier, of course, one is obliged to conclude that my voluntary decision made at that time was in fact determined by the circumstances in which I had been placed prior to that. There is both free will and determinism, isn't there? Recently, I've been using the term "open world view" to refer to the idea that one is living in a world with the kind of unknown element that I spoke of earlier. But if one tries to trace things back to a period before one can remember one never gets anywhere, so I'll start around the time when I first felt that I wanted to devote myself to learning: that is, the period from high school through university, when the resolve—the will—to do research into physics had become independent of other things. When I say "other," I know that a human being naturally thinks of and does all kinds of other things; one isn't only studying from one end of the year to the other. However absorbed one may be in one's studies, one gets up in the morning and brushes one's teeth and washes. This fact in itself is unrelated to the fact that one is studying a certain branch of learning for all one is worth—un-related, that is, in the sense that there are things that everyone, scholar or not, does. There are all kinds of such things in human life. But in my case, I feel that there was, in addition, a strong tendency for the pursuit of learning to control my other thoughts and actions. This is shown specifically in some of my old diaries, written during a certain period in my youth, that I still have. They deal with various topics. And sometimes they're not written up continously as diaries but are simply memos and random thoughts jotted down on sheets of paper.

INOUE: You mean, made by yourself?

YUKAWA: Yes. I can only find those written after I left university—the ones before that have been lost—but I find that these diaries, memos, and the like are full of rules for my own behavior—instructions to myself to approach things in this or that way. It shows more clearly still in the daily schedules I set up for myself. I was to get up every morning at such a time, and do this from such a time to such a time, and do this in the afternoon. Many of the entries lay down some sort of daily timetable.

INOUE: Was it the same even when you were a student?

YUKAWA: I'm sure it was. But to lay down a schedule is different from putting it into practice. Often you don't keep exactly to the times, or you miss doing something that you were supposed to do every day. So after six months or so, you have to draw up a fresh schedule for yourself. And each time it's a little less strict than before. But though my schedules may have changed, the fact that I made them so often shows how tight a rein I kept on myself, doesn't it? The only question is whether I really lived up to the rules I laid down for myself; personally, my feeling is that I succeeded only about fifty or sixty percent. Either way, the fact remains that here was a human being who aimed to become a scholar; then, after starting on his studies, he strove for ten years or so to devote the whole of his life to a particular aim as a research worker, or at least to maintain a continuous progress. That approach and my present "open" view of life, though perhaps not incompatible, are at least, I feel, somewhat different.

My Personality

YUKAWA: To go back still farther, from my middle school days to my middle years at high school an extremely important factor that made me determine to go in for physics, and for theoretical physics in particular, involved my personal ability and inclination —you could lump them together as aptitude. So that I could decide on the course I should take, these seem to have been under consideration from all kinds of angles in my subconscious mind over a long period. Here, I seem to have been strongly influenced by the desire to be able to carry on my studies without being bothered by anything outside my actual research.

INOUE: I believe you once wrote that you analyzed yourself and concluded that you weren't good at giving expression to yourself.

YUKAWA: What it comes down to is that not being very sociable by nature—because I'm poor at dealing with other people— I wanted to work in some field of study where there was little necessity for such things. Not, of course, that theoretical physics is the only study that answers that requirement.

INOUE: Since your father[2] was involved in all kinds of different things in his work, I imagine that so far as the external conditions, or influences, were concerned there were all kinds of possibilities open to you.

YUKAWA: So far as anything inherited from, or any influences from my father are concerned, or any special hopes he had of me, they did not fit in very well with what I've just said. My father was a geologist, a geographer, and many other things. In a broad sense, he was a natural scientist I suppose, but as such his studies brought him into relatively frequent contact with human society.

As a young man, you see, he worked in a geological survey center, and had to travel all over the place preparing geological maps of a large part of the country. Since a geologist has, of course, to survey in order to make his maps, inevitably he must go into the mountains and to places that are inaccessible, where he has to find lodgings and therefore comes into contact with all kinds of people. It involves a considerable amount of physical strain and, I'm sure, even a certain amount of danger.

So my father had to have a lot of contact with other human beings and he often had to go somewhere completely unfamiliar and find an inn to put him up. It's not the kind of study you can carry out in isolation from the world. This was even more true of geography, since my father was less interested in natural geography than in anthropogeography. Anthropogeography, in short, is a branch of study that's very much involved with people. Its purpose is to consider the interrelationships between human society, human life, and the natural environment in which human society finds itself. From the human standpoint, there's a kind of feedback and also a kind of adaptation. That's the kind of problem it concerns itself with. Basically it is the kind of study that needs to maintain the closest kind of exchange with human beings. But that was precisely what I found most trying. So I did not fall in with my father's suggestions that I take up geology or geography.

Engineering also, for the same reasons I've just given, was most unsuitable. My eldest brother took up metallurgy. But here again, he was always going into the mines and meeting people connected with mining or metallurgical enterprises. I'm just not suitable for that kind of thing. And another thing—to put it mundanely—I'm not good at haggling over the price of machines and the like. All this meant, unfortunately, that the things left that I could do were gradually reduced. In the natural sciences, you see, you normally have to do various kinds of experiments involving various kinds of apparatus and machines. That means that you have to buy this equipment somewhere when you need it. To buy it, you have to deal with the people who sell it. By dealing, I don't just mean bargaining over the price, but giving instructions about the type of apparatus, the time when it's required, and so on and so on. In my college days, that kind of thing seemed a fearful trial for me; in fact, it appeared quite impossible.

Another reason was that, generally speaking, I'm extremely clumsy. The average Japanese is very clever with his hands and very adept at mastering special skills. The Japanese, in fact, may well be the most dexterous nation in the world. I had a very strong sense of inferiority towards other people of the same age where manual skills were concerned. I could give all kinds of examples—mechanical drawing or glasswork, for instance—but the important thing was that I was no good at experiments where one needed to be clever with one's hands. What did this leave, then? If one looked outside the field of natural science, there were plenty of things. Even eliminating them one by one according to the method I've just used, it still left the cultural and social sciences. In the social sciences, however, I had no interest at all. In fact, this is very much connected with what I have already said, because I was entirely unconcerned about the kind of society I was living in, or the abstract structure of human society, or the nature of man's economic life, or the way society or politics worked. In short, to consider such questions, even when the considering was done in one's own study or at a desk at the university, was, in itself, to entertain indirect relations with society, and I had little taste for them, therefore. That, I believe, is how my mind worked.

Reading in My Middle School Days

INOUE: It was a period when all kinds of things were going on in society, wasn't it? You were born in 1907 and left college in 1929, by which time the socialist movement had already taken hold to a certain extent. Do you think perhaps that these feelings of yours about the social sciences were fashioned at a still earlier period in your life?

YUKAWA: Well, now . . . to go back to my middle school days, which means from 1919 to 1923, there was quite a vogue for "philosophies of life" around the time I first entered middle school. A typical case was Tolstoy's philosophy. Some of the young people at that time prided themselves on being Tolstoyans. One of my contemporaries at middle school actually joined Saneatsu Mushakoji's New Village. My elder brother Shigeki Kaizuka[3] who had heard about Tolstoy from a classmate, talked about him a great deal. So I began reading Tolstoy. This was not long after I'd entered middle school, an age when one starts pondering questions such as "What is life?" It is a time, too, when one begins to have doubts about all kinds of things.

INOUE: Dr. Kaizuka once remarked that he himself liked Tolstoy while you preferred Dostoevsky—would that have been around the same time?

YUKAWA: No, that came a little later. At that time I still knew nothing about Dostoevsky. Nor had I read any of Tolstoy's novels, even. There was a slim volume called *On Life*, though. My brother had read it and knew all about it. So not to be out-done, I read it too. And I was impressed in a way.

INOUE: Impressed. . . ?

YUKAWA: In a very simple kind of way, of course. It was after that that I began to read Tolstoy's novels. The kind of novels I read in my middle school days were, for example, those of Soseki Natsume[4] just as everybody did and still does. In that respect I was no different from anyone else. What followed, though, was a little different; in my later years at middle school, in my third or fourth grade, I'd say, there were some in the same class who were reading social science—which meant, in short, Marx. They were about the same age as myself, and I was vastly impressed by their precocity. Tolstoy's philosophy had relevance for me in relation to my own life. It might relate to social problems, yet since its basic relevance was to my own life, I could understand it. But when it came to social science, I understood little and cared less. This was the middle of the Taisho period, around 1920, so there were a lot of university students who studied social science and Marxism. It must have been around then that the Shinjinkai[5] was flourishing at Tokyo Imperial University, too, though I didn't know much about it. Anyway, there were people of my own age who were remarkably sophisticated in that respect, while I was completely indifferent.

What direction, then, did my interest take? As I've written elsewhere, I acquired a taste for the kind of fatalistic outlook represented by Laotse (老子, 604?–531? B.C.) and Chuangtse (荘子, 4–3 century B.C.). I also read the immensely nihilistic novels of Hakucho Masamune (1879–1962) and was attracted by them.

Anyway, one thing that was consistent throughout was that my state of mind was extremely shut-in. I was very isolated. Being isolated, I found the world annoying, which in turn produced a kind of fatalistic pessimism. And I thoroughly enjoyed it at the same time. It was a kind of sentimentalism, I imagine. At the same period I also read the works of such novelists as Genjiro Yoshida (1886–1956) and Hyakuzo Kurata (1891–1943). I was very much attracted—perhaps what I enjoyed was soaking up the melancholy mood they evoked.

The Road to Physics

Mathematics

YUKAWA: So much for attitudes to life, then. Perhaps I should say something now about how this connects up with scholarship.

At middle school, I was relatively fond of mathematics, the reason being that one could solve a problem by oneself, without relation to other people. The process of solving it was also extremely enjoyable in itself. Around that time a book called *Geometry Made Easy* was published. It was very interestingly written, rather like a history of mathematics in parts. And it had all kinds of exercises, the methods of solving which it explained in an interesting fashion. I used to read such books and spend hour after hour absorbed in working out geometrical problems. I might mention here, to digress a little, that around that time there were a whole succession of works with titles—one might be talking about the present day—such as *How to Study Algebra*, and about every other subject you could think of. Their aim, in fact, was to prepare one for entrance examinations. If you worked through them, you stood a better chance of passing the examinations to high school. I didn't read them much, having no particular interest, but the point is that such books were popular at the time. In those days, the entrance examination to high school was the stiffest hurdle in the educational race. Even then, people to some extent made special preparations for the entrance exams, though it was nothing like nowadays. I myself did the same to some extent. But in mathematics I was confident and didn't worry about the entrance exams much.

The Path to Physics

YUKAWA: To get back to the main subject, then: the choice of futures gradually narrowed for me. The process was one of deciding that I wasn't suited to first this and then that, one subject being discarded after another. But the humanities had still not been discarded. I was still left the possibility of, for example, philosophy, literature, or history. Geography may be included in history in the broad sense, but it did not attract me for the reason already mentioned. I always liked history, and still do, but I couldn't help feeling that as a subject for research it was too amorphous. I sensed vaguely that even if I were to take it up, I would never know when I had got a real grasp of it.

I was extremely fond of literature, too, and still am, but again it was too vague for me to imagine making it my life's work. I felt it preferable to treat it as a pastime.

Also, mathematics still remained, but on entering high school I felt I would give up mathematics. The most essential reason here, I feel, was that though I was extremely shut-in and felt I wanted to do something that would not be too closely connected with the world, anything that was too remote from reality—I feel that's the best word here, whatever the precise meaning—was empty and meaningless; or if that's putting it too strongly, that unless a thing was somehow connected up with reality it was, if not actually empty and meaningless, at least difficult to be wholeheartedly satisfied with in one's studies. There were all kinds of superficial reasons, but the real one, I think, lay here.

Encounter with Quantum Theory

YUKAWA: Next, then, philosophy. It was 1923 when I entered the Third High School. In no time at all I was going regularly to the school library to read. There were all kinds of books in the collection. At that time, a philosophical series was already in process of publication by the firm of Iwanami. Quite a few volumes had already appeared, I think. Among them was *The Natural Sciences in Recent Years* by Hajime Tanabe (1885–1962). There was also a single volume entitled *An Introduction to Science* by him. I found them extremely interesting. One of the few prolific writers among scientists at the time was Jun Ishihara (1881–1947), who was represented by his *Theory of Relativity* and *Fundamental Problems in Physics*. So my interest began to be directed towards works concerned primarily with the philosophy of science rather than works on philosophy proper.

Among the things explained in Hajime Tanabe's works was quantum theory. However many times I might read it, though, it made no sense to me. This very incomprehensibility was an enormous attraction, and I had a strong feeling at the time that this was something I ought to study. Ishihara's account of the theory of relativity I understood very well. Well, that's putting it too strongly, but I understood it more or less. There's nothing particularly odd about relativity. Quantum theory, on the other hand, is a most peculiar affair. So I thought it might be a good idea to study it.

INOUE: Was it around that time that Einstein came to Japan?

YUKAWA: That was when I was in the fourth grade at middle school. It was just before I entered the Third High School, at a time when I had not yet made up my mind to become a physicist. So all I knew was that a terribly distinguished scientist called Einstein who had thought up some very remarkable theory of relativity had come to Japan. That was the whole extent of my awareness. There was a certain amount of aspiration too, I suppose. So when I entered high school and read Tanabe's and Ishihara's books, I was fascinated by what seemed to be a very new type of physics in the making.

The High School Student under the Old System

YUKAWA: The high school student in the old days, though, was a happy-go-lucky fellow. He went in for all kinds of sports and became a member of one of the organized cheering sections for baseball games. On anniversary days, the pupils of the Third High School would work hard setting up—what were they called now? Stalls? Eating places? In tents. . . .

INOUE: Refreshment booths.

YUKAWA: That's right. They set up refreshment booths, where they served coffee and doubtful eatables. The students of a class were obliged to sell tickets in advance for their class's refresh-

ment booth. They went out selling them in the streets. I myself
went with them once, and I still remember how unhurried
Kyoto was in those days. When there was an anniversary day,
crowds of ordinary townsfolk would turn up and eat and drink
the varied and dubious snacks on sale at these stalls. I shudder now
to think what the state of sanitation must have been like, but at
least there were no outbreaks of food poisoning. I imagine the
bacteria must have been as easygoing as everybody else . . . the
dormitories, too, were opened to the public. They were pretty
squalid, but every room would spend a lot of thought on what
were known as "decorations."

INOUE: "Decorations," indeed.

YUKAWA: Yes—you'd expect them to be attractive, but they
deliberately made them messy. Nowadays young people in Japan
use the word *kakkō ii* ("smart" or "stylish") a lot—everything
they wear or have about them has to be *kakkō ii*, but in those days
most high school boys tried to make themselves look as scruffy
as possible. Scruffiness in itself was a goal to be striven after. It was
no good looking only half a mess. The caps the boys at Third
High School wore were a good example. One would think new
students would be proud of their nice new caps, but in fact they
would deliberately get the older boys to let them have their old
caps, or they would soak their new ones in water to make them
dirty; everything about them had to be messy. When the young
people of today talk about being "smart," they may mean just the
opposite, and the students of those days, too, had their own way
of being "smart." I myself, though, lived at home and was different from those who lived in the dormitory; I didn't get much
involved in the general trend to scruffiness.

Ignorance of the Social Sciences

YUKAWA: All this reminds me, too, of how innocent of the
world I was. I went to school, where I attended classes, studied in

the library, and even played a certain amount of sport, but of
other things I knew almost nothing. The year before I entered the
Third High School, the students there went on strike. The cause
was the firing of a number of teachers by a new principal called
Kaneko. It was rather like the Sawayanagi affair at Kyoto University, when a number of professors were discharged on grounds
of incompetence. The students went on strike out of sympathy
for the dismissed teachers. My elder brother Kaizuka got involved
in it shortly after he entered the school, and with the others he
barricaded himself in the dormitory for a while. My father, who
was worried about him, went in the evening to fetch him home,
taking me along with him. When we reached the entrance to the
school on Higashi Oji street, the gate was shut. Father tried to
find some way of getting in. A student who was inside, looking
over the top of the gate, insisted that we couldn't come in. Although my father was very different from me, having a loud voice
and a determined manner, he failed to get his way with the
students and had to give up in the end. So I went home with him.
But at that time I had no idea at all of the significance of what we
were doing. All I knew was that I had gone along with my
father; that my elder brother seemed to be inside; that we had
not been allowed in; that because we were not allowed in I had
come home again with my father. I may have known that a
"strike" was on, but as to what a strike was, or what they were
striking for, I knew nothing. It's odd, isn't it? I had absolutely
no understanding of, or interest in, such things.

Another thing: at Third High School, there was a class on
economics and law, but what the teacher said went in one ear and
out of the other. I wonder how I ever got through the examinations. I must have studied a little just before the exams and got
something into my head, for I got through all right, but I have
no recollection of what I wrote or of what the teacher told us.
In the other classes, I understood what was going on, but not in

economics and law. There were lots of subjects I didn't enjoy, but this was a little extreme. So even today I'm still like a primary schoolboy where social science is concerned. Nowadays there's a social course even at primary school, I believe? But even after I entered high school, I still understood nothing at all of social science, law, or economics. Or rather, it was less a question of not understanding than of feeling that such things were quite irrelevant to me from the very beginning.

The Choice of Physics

YUKAWA: I've digressed considerably, but, to get back to the main subject, one of the things that pleased me at high school was the fact that I started learning German as a second foreign language. It delighted me to be able to read German a little, and I conceived the ambition to read a book in German. It must have been when I was in third grade that I found in a bookshop the first volume—on dynamics—of Max Planck's *An Introduction to Theoretical Physics*. It was not an attractively bound book, but I was very pleased at having found it all by myself. Nobody had told me about it; I had gone to a bookshop and there it was. I bought it and took it home intending to read it. When I did so, I found I understood more than I might have expected. I knew Planck was a very great physicist, and the book was based on lectures he had given at Berlin University. I found it fascinating. The thought was straightforward and logically reasoned, easy to absorb for a simple, unsophisticated brain such as mine. I was grateful that such a well-known scholar should write something so intelligible to a high school boy like myself. What made it all the more exciting was that I also knew, from the books by Tanabe and Ishihara that I had already read, that Planck had been the originator of quantum theory. And it strengthened the feeling that I should take up physics myself. So I have the feeling that among the things that had the greatest influence on my decision to take up theoretical physics were the works of Tanabe and Ishihara, followed by Planck's book. In this way, I finally applied for admission to the Department of Physics at Kyoto University. There was an entrance examination at the time, but it presented little difficulty. From then on I envisaged clearly what I meant to do. Not once had anyone told me to become a physicist. I had chosen my own course—which was a very good thing for me.

INOUE: Didn't your father try to make you give it up?

YUKAWA: No, not to give it up. Around the time when I started reading Planck my father urged me to read a book in English on geology, hoping to get me to take the subject up, but I didn't show much enthusiasm for it so he didn't try to persuade me any further. Nor did he oppose my taking the physics course. So my entry into the physics department at Kyoto University went ahead without a hitch.

One day shortly after my application had been accepted my father came home and told me he had met Professor Naito[6] and that when he had heard I was going into physics he had greatly approved of my taking up what he called a "basic branch of learning." I was highly delighted. Konan Naito was a close friend of my father's; though I had not met him myself, he cropped up frequently in my father's conversation, and I had the idea that he was a particularly great scholar. I was overjoyed that such a great man should have said such a thing; it may seem a very trifling occurrence, but it encouraged me greatly.

INOUE: You entered university in 1926, I believe. At that time the university lectures would have included no mention as yet of the quantum theory, would they?

YUKAWA: No. There was not much of what you could call quantum theory in the lectures of those days. There was a brief description of it in the course on the theory of heat, but just at that time a more up-to-date quantum mechanics appeared in Europe and was gradually receiving widespread acceptance.

Toward Meson Theory

INOUE: Your period at college coincided with the dawn, as it were, of the quantum theory in Japan, didn't it? Perhaps you could say something about the period from then up to the formulation of your theory of the meson?

YUKAWA: I've written many times about the course of my thought until I reached meson theory, so I'll make it brief. While I was at high school I felt that the quantum theory was extremely fascinating and revolutionary and that I wanted to study it intensively. At college, however, I found that what I had been thinking of as the quantum theory was already out of date, and that a more advanced quantum mechanics had appeared and was causing a great upheaval in the world of physics in Europe. Anyway, I resolved that I must study it. I did so, while all the time quantum mechanics went on being applied to all kinds of fields with a steady succession of successes. Germany was in the lead, but, in the small countries round about—especially Denmark, Holland, and Switzerland—and in England, France, and Italy, young scientists were pressing ahead with remarkable results. I was rather dismayed. I began increasingly to feel that I had become a scientist rather too late. So I debated with myself what I should do on graduating; but I no longer felt like taking up any other branch of study. My determination to study theoretical physics was unchanged; the question was what aspect of it. After much thought I realized that almost nothing was known about the atomic nucleus and cosmic rays and also that there were very few people working in the field. So shortly after graduating from college I decided definitely that there was no alternative but to go in that direction.

INOUE: At that time, there were also the problems of applying quantum mechanics to the structure of the atom and to the molecule, weren't there?

YUKAWA: Yes, there were many problems of that kind, but they were being cleared up at a great rate. There seemed to be little hope of catching up. I felt I was rather too late there and I decided to go into the field of the atomic nucleus and cosmic rays, which was something that other people had not paid much attention to yet. Various other things happened in the meantime, but about three years later the discovery of the neutron made me feel definitely that I should have to get moving if I was to do research into the atomic nucleus. I feel things were different then from nowadays—in Japan just as elsewhere. It was considered, if anything, normal for someone to produce some concrete achievement, whatever its relative importance, during their first two or three years after graduation. I had a sense of urgency, but, however pressed I felt, my work seemed to make no progress.

I'm an extreme kind of person; I can never work on a problem that I've been told to solve by somebody else. My subconscious always rebels against being ordered to do something. Personally, I look on myself as a docile kind of man, but my subconscious seems to be a rather nastier character. Once it happened that Professor Yoshikatsu Sugiura (1895–) came to give lectures at Kyoto University and he suggested that I take up a certain problem. I thought I'd have a go at it, but my subconscious said no. So I tried to find some excuse for getting out of it. The question involved was to find a plausible explanation for the puzzling results of a certain experiment. I myself suspected that since the results were odd, there was something wrong with the experiment itself. So I read the scientific journal in which the paper on the experiment had appeared and, as I had suspected, found something wrong with it. Since the experiment itself was dubious, I decided to do nothing about the explanation. Eventually, it became clear that the experiment itself had been at fault. It was lucky that I had not done the work. I can't help feeling there's something perverse about me. My idea of studying the atomic nucleus and cosmic rays suited me, therefore, since few other people were doing it and since I had not been told to do it by anybody. Soon afterwards I changed to Osaka University and worked at something or other by myself but didn't really get anywhere. Nor could I have been expected to. This kind of thing went on until one day I hit on the idea of the meson.

Recollections of Joint Research Work

YUKAWA: This was in the autumn of 1934, some five years after I'd graduated from university. In those days, five years was a long time to spend in producing a thesis. I imagine I was considered to be stupid for taking so long to produce results from my work. Well, perhaps I was. . . . Subsequently, Shoichi Sakata (1911–70) and Mitsuo Taketani (1911–) joined me in my research. A little later still, Minoru Kobayashi (1908–) joined us, and we worked together developing meson theory. We'd more or less completed the first stage of our work by 1938–39. In the meantime, similar kinds of research had got going in England and a few other countries. I had gone about as far along that line as I thought possible when I was invited to an international conference in Europe and went abroad for the first time in my life. That was in 1939.

INOUE: Just at that period (1936–38), I was a student at the Third High School. I remember reading abstracts of papers by you and your collaborators in the Japanese science magazine *Kagaku* put out by Iwanami. I heard later that the summaries were made by Professor Yoshio Fujioka (1903–). Of course, I failed completely to understand them, but I remember being greatly attracted by one title "On the Interaction of Elementary Particles." You came back to Kyoto University in 1939, the year I went there from high school, didn't you?

YUKAWA: Actually, it's a little more complicated than that. The year before, Professor Kajuro Tamaki (1886–1938), my teacher at Kyoto University, had died, and about one year later it was decided that I should go back to Kyoto to succeed him. Then, just after that, I was approached by professors in Tokyo University and asked to be a professor there. They were so very enthusiastic that, even though I already had an arrangement with Kyoto, it was difficult for me to turn down the offer flatly. Being very much in two minds, in the end I compromised by getting them to agree to let me lecture at both places concurrently, so that from 1941 I was giving lectures from time to time at Tokyo University. Unfortunately, however, the air bombardment got steadily worse. In 1943 it got too much for me to keep going backwards and forwards to Tokyo, so in the end I asked to be released and settled down to work in Kyoto.

Such being the situation, I had a room in the Physics Department of Tokyo University as well, where I spent a little of my time. Around that time, Kunihiko Kodaira (1915–), the mathematician, was very interested in physics. It was suggested that since he had a good brain we should get him to study physics, and at one stage I, too, tried to persuade him to be a physicist. At the time, he was keenly studying the S matrix theory which had been proposed by Heisenberg, and I thought it would be a good subject for a man like Kodaira. In the end, however, he never became a physicist, making his name instead as a mathematician. That is excellent, of course, but considering the great developments that the S matrix theory underwent in the sixties, I can't help feeling that his presence was missed in the field of physics.

INOUE: To change the subject, you went to Europe in 1939 on an invitation to attend the Solvay conference, didn't you? The conference was cancelled on account of the war and you came back to Japan, but it was the papers to be read at the conference, and in particular those by Heisenberg and Pauli, that in a sense provided the starting point for our studies in your seminar, wasn't it?

YUKAWA: Yes. As I mentioned a while ago, I was invited (by

the Solvay Institute) just at the time, in 1939, when I moved back
to Kyoto University from Osaka. I went to Europe to attend the
conference and installed myself first of all in Berlin. But almost
immediately Germany went to war, so I got out and came back
to Japan via America. Two years later, Japan herself went to war,
but around 1942, while the Pacific war was still in its early days,
we were still able to study comparatively undisturbed. Kyoto in
particular was one of the quietest places. It was around this time
that Shoichi Sakata and Yasutaka Tanikawa (1916–) first put
forward their two-meson theory. Shortly after that, Sakata moved
from Kyoto University to Nagoya University.

INOUE: That was around the summer of 1942, wasn't it? I went
to Nagoya too, together with Professor Sakata. The university
was only recently established then.

YUKAWA: You and Seitaro Nakamura (1913–) contributed
to the development of the two-meson theory. At Kyoto University during that first period, things were still relaxed enough
for us to go from Kyoto to Nara for hiking on our days off; it
was no longer peacetime but the war had only just begun, and
the outlook was favorable for Japan. In our room we had Sakata,
Tanikawa, Nakamura, yourself, and a number of others, but only
around ten in all. And one day, somewhere in Nara—Tobihino,
perhaps. . . .

INOUE: Near the Nara Hotel.

YUKAWA: Where was it now, near the Sarusawa Pond? Anyway, we were sitting on the grass in that area talking of this and
that when Sakata and Tanikawa began to talk about the two-
meson theory. After a lot of discussion, we decided that it was
worth investigating further. That was virtually the beginning of
the two-meson theory.

INOUE: I remember that I had an argument with Professor
Sakata concerning the two-meson theory on the train to Nara.
We were both standing. . . .

YUKAWA: In those days, pennants on top of poles were used as air-raid alerts. I remember catching sight of what looked like pennants. Startled, I looked again and found they were the multicolored carp banners put out for the Boys' Festival on 5 May. I was relieved and went on with the conversation. So it must have been around the beginning of May.

I was interested in the two-meson theory at the time, so I told them to work hard at it, but at the same time I myself was beginning to think along other, somewhat different lines. The two-meson theory alone would, I felt, not explain everything; another, even more basic problem still remained, which I doubted could be explained within the existing framework of relativistic quantum field theory. It would be necessary to think farther ahead, and I had begun at various meetings to draw circles on the blackboard.

INOUE: It was around the same time that Oppenheimer, in concert with Schwinger and others, began to attack the prevailing meson theory, wasn't it? They complained of a discrepancy where the lifetime of the meson was concerned.

YUKAWA: It was the same with us; Sakata calculated the life of the meson and found the discrepancy through his experiments. That was one of the things that led to the two-meson theory.

INOUE: The views put forward by Oppenheimer and his associates were pessimistic—in a sense, a kind of challenge to us—since they pointed out the discrepancy between the theoretical and observed life and collision cross section of the meson, and on those grounds they suggested that in future the meson theory would suffer the same kind of fate as had overtaken the ether theory of light when it was replaced by the theory of relativity. I remember even now how their paper aroused my fighting instinct. I remember saying to Professor Sakata at the time that in view of the discovery of the meson in the cloud chamber and the undeniable achievement of Yukawa theory in grasping mutual transformations between elementary particles, I could not accept their outlook as correct, and that there must be a way out somewhere. I couldn't help feeling a slight touch of gratification a few years ago, at the ceremony celebrating the thirtieth anniversary of the meson theory, when the passage in Oppenheimer's congratulatory telegram was read out in which he more or less apologized for the confusion caused by his own misjudgment in the early stages of the history of meson theory.

YUKAWA: There were two possible approaches there. One was to ascribe all the difficulties involved in the meson theory to the inadequacy of field theory as a whole. The other was to retain the meson theory and to overcome at least some of the difficulties by improving it. The two-meson theory offered a very good means to the second approach.

Tomonaga's Approach

INOUE: The idea was to come to the rescue of meson theory and, since field theory could not succeed all at once, to combine both approaches, wasn't it?

YUKAWA: The policy of our group was to pursue, on the one hand, the two-meson theory, so far as it went; on the other hand, to deal with more basic problems. Although I was aware that it was impossible to solve them in a hurry, I myself wanted to tackle them. That was why, among other things, I drew circles on the blackboard. There were symposia on meson theory at the time, and I drew my circles at these meetings in both Kyoto and Tokyo. I embodied my ideas in a paper in question-and-answer form entitled "On the Basis of Field Theory," which was published in installments in the science magazine Kagaku. Professor Shinichiro Tomonaga (1906-) was sceptical about the circles that I was forever drawing and in 1943 put forward what he called "super-many-time theory." I was very interested and impressed. Professor Tomonaga applied it to quantum electrodynamics with great success. I for my part had not, even so, given up my circles, but soon the war situation grew worse and it gradually became difficult to settle down to study. . . . After the end of the war, Minoru Kobayashi and I got together and decided to put out a journal named *The Progress of Theoretical Physics*. I would like to leave a detailed acount of what we have done for another occasion; the point is that even then, after the end of the war, I was still hoping obstinately to do something with my long-cherished idea. At that time, however, I was asked by Oppenheimer whether I wouldn't go to America. I went, with the assurance that I could really settle down to my research work at the Institute for Advanced Study in Princeton without having to give any lectures. While I was there, I conceived the idea of non-local fields and wrote various papers about it. Even so, that was not the end. When I came back from America in 1953, I went ahead with my studies at the Research Institute for Fundamental Physics, newly established at Kyoto University. As time passed, things gradually developed beyond nonlocal fields, and recently I've been considering the idea of what I call "elementary domain," but I don't know how things will turn out from now on. . . .

Toward a Unified View of the World

The Attachment to Learning

YUKAWA: So what I'm doing now can be traced right back to 1941 or 1942, which shows how persistent I've been. If you trace it still further, it goes way back to a paper by Heisenberg and Pauli on quantum mechanics of wave fields that appeared around the time I left university. It was a crucial piece of work and I read it over and over again, but it left very fundamental problems unsolved. I decided I would solve them myself. But there was little likelihood of achieving such an inordinate ambition when I was only just out of college. So I put it aside for the time being and went into the theory of the atomic nucleus and cosmic rays, which led me to meson theory. Nevertheless, I had not forgotten my early ambition at all. From the time I graduated and left college right up to this day, I've had the idea of building up some theory that has firm foundations. It's been nearly forty years now, but I'm still at it. I shall retire from my professorship in something over two years' time. The retirement age at Kyoto University is sixty-three, and I am going to retire in March of the academic year in which I reach that age. Of course, it doesn't really make any difference whether I am at the university or not, but one feels it as a kind of demarcation line all the same. It's a very good thing that there should be such dividing lines, and I want if possible to wind things up more or less during those two years. I don't know whether things will go as I hope or not.

Looked at in one way, you know, human beings go on doing the same thing for thirty or forty years. It's not so very different from the way one gets up and brushes one's teeth every morning. In brushing one's teeth, there's no progress. In scholastic studies, there ought to be progress, but in practice that doesn't happen so often. It may well be that one spends thirty years without achieving anything. Even so, I want to try. Another two years, and it will be goodbye to university. I shall be more than pleased if I manage to get my ideas into shape by then. If not, then I shall carry on further. If they're in shape by then I shall quit physics with good grace. There are any number of other things I like and want to do. Having lived for longer than sixty years, I've lost the narrow idea I had at first, that the study of theoretical physics is my only vocation. There are plenty of other things I feel interested in; whether as a serious study or just as a pastime, I couldn't say yet. It might well be as a pastime; there's no particular need to make a serious study. Poetry would do, or philosophy, or history, or I might even feel like a belated sally into the social sciences. By now I have the feeling that political science, law, and economics would all be rather interesting. In this way, unlike my youth, I'm interested in all kinds of things, and from now on I may gradually become unable to tell what my main profession is. Or rather, I suspect that I shall stop having a regular profession— shall be freed from it so that I can study more widely for my own pleasure. The farther I go back in time, the more the reverse was true: for a while at twenty I became very narrow in my outlook and had a very strong sense that this was the one thing that I was born for, that this was what I would do. One does something because it's the only thing one can do.

INOUE: You've often said it's a kind of obsession.

YUKAWA: It is, undoubtedly. In the *Analects* of Confucius (551?– 479? B.C.) it says that the scholar of old used to study for his own sake, whereas the scholar of today studies for the sake of other people. A wry kind of remark. I for one certainly do it for my own sake. And yet, if one studies for many years and acquires a certain amount of recognition, you have to do it for other people too. You have to consider society as well. Although this is what Confucius said, in practice he did a lot for other people. I don't think I was ever antisocial, but I was certainly asocial. Nowadays, though, I can no longer be that. That's what makes my life difficult.

The Development of Science and the Purpose of Life

INOUE: To finish up, then, I wonder if there's anything you'd care to add concerning future trends in physics?

YUKAWA: As I said earlier, Planck was a great scholar. It was he who in 1900 hit on the quantum theory, an event that marked the true beginning of the twentieth-century revolution in physics. In 1908, when he was professor at Berlin University, Planck was invited to speak at Leyden University in Holland, where he gave a lecture of lasting significance. I believe it to be the most outstanding of his lectures. In it, he lays great stress on the idea that physics is striving to find unity in reality by transcending, by getting away from the human senses—trying, in short, to discover some unified image of the world of physics. He talks about it passionately. I was deeply impressed when, as a university student, I read this lecture in a collection of Planck's lectures, and however often I read it the impact is revived.

In one sense, things have turned out as he said. In short, physics has shown a progressive tendency to divorce itself from the human senses, or should I say from humanity in general? It has become progressively detached from mankind. That is as Planck foresaw. But what, on the other hand, of the unified picture of the world that was Planck's ideal? As it is today, physics is an enormous distance from that ideal. A tremendous number of learned papers on physics has been published since his day, yet physics nowadays is lacking in effective attempts to widen and deepen its perception of nature (a phrase I'm fond of using) and to find some unity in it. Another thing is that the expectation that physics would become separated from humanity, from the human senses, of which Planck spoke has proved to be true, and it has made it more difficult to construct a unified world picture. Twenty-odd years after Planck, quantum mechanics came into being. From it the so-called theory of observation has emerged. The problem here is as follows: the facts that we learn directly and indirectly via our sense organs on the one hand and our theoretical understanding of nature in terms of mathematical laws on the other— the two things I customarily refer to as "facts" and "laws"—cannot easily be united into one and the same thing. Unification of the world picture of physics in the naïve sense in which Planck envisaged it is not so easily achieved. That, in short, is the state of affairs that has come about since the appearance of quantum mechanics.

It's not yet possible to tell what effect advances in the theory of elementary particles may have on the second of the two things I've just mentioned, but the first—the alienation from the human senses, which was still not such a great problem in Planck's time—has become far more troublesome recently, on account of a fearful one-way trend towards abstraction. Terms such as "facts" and "laws" sound harmless enough, but in practice laws nowadays seem to be confined to formidably abstract mathematical formulae. The facts themselves, too, have increasingly become facts that are only realized under certain very special circumstances. Take cosmic rays, for example—they contain certain particles that are extremely few in number but have enormous energy. Such things have become important because it's in them that one is liable to discover new facts. Alternatively one builds a huge particle accelerator. The new facts discovered through its use then assume great importance. Either way, we are getting farther and farther away from the human world. Actually, there are not so many new facts. You come across them only rarely. Only the physicists make great efforts to find them and compare them with the theories to test whether the latter are correct or not. And these theories have become so terribly abstract, so abstract that one begins to have doubts about their significance. If one is concerned exclusively with such things, then for what purpose precisely is one engaged in theoretical physics, in fundamental physics? When one turns back to one's own views on life and endeavors to relate physics to some sense of purpose, one begins to have the gravest doubts. Indeed, one wonders whether they have any relevance to the purpose of life at all. This was what prompted Planck, at the beginning of the twentieth century, to set forth his very simple and basic convictions. Subsequent physics did not take the form that he expected, but even today I can perfectly understand that simple and naïve desire of his for a unified view of the world. To talk of it is easier than to realize it, but to strive

towards such a goal is surely, even today, the most essential
justification for physics. To set up increasingly massive equipment, or to use ever more efficient computers has no significance
in itself. The essential significance lies not in these things themselves, but in the attempt to expand and deepen our view of the
world—to put it differently, to get back to natural philosophy,
to get at the innermost depths of the mind. Without that, my
attachment to physics is pointless. It is precisely because that may
be possible that I am studying physics. Unfortunately, I may never
be able to reach that point. Resignation may be important for
human beings; the time may come when I have to give up. But
I have not given up yet.

Notes

1. Takeshi Inoue (1921–). A physicist. Professor in Kyoto University.
2. Takuji Ogawa (1870–1941). A geologist and geographer. Late professor in Kyoto University.
3. Shigeki Kaizuka (1904–). An Oriental historian. Late professor in Kyoto University.
4. Soseki Natsume (1867–1916). A great novelist and scholar of English litera-ture in the Meiji era.
5. Shinjinkai. A socialist students' group founded in Tokyo University in 1918.
6. Konan Naito (1866–1934). An Oriental historian. Late professor in Kyoto University.

II ON WAYS OF THINKING

1 East and West [1952]

It is just four years since I went to America. I realize, looking back, that Japan and the world have both changed a great deal in the meantime. Even compared with two years ago, when I last came back to Japan, I have the illusion that a very long period of time has passed. Yet there has been no particular change in our daily lives since we started living in America. My apartment was only two or three minutes away from Colombia University where I have been professor since 1949, and apart from the time spent daily going to and fro between the two, I had relatively little contact with the town, so that I tended to be isolated from the average American as well. There was, as ever, a steady flow of visitors from Japan, though the spate of that flow gradually diminished.

Yet although there was no marked change in the externals of everyday life, there was a considerable change in my own state of mind. If I compare my first two years and my second two years, the former was, in a sense, a period in which I was experiencing the truth of my long-held conviction that human beings, whether they were Japanese, American, or what have you, had common human qualities that made it possible for them to understand each other and live pleasantly together. The second two-year period brought no change whatsoever in that conviction, yet one further question now came to impose itself insistently on my awareness.

This problem was my increasingly strong doubt, the better I got to know America, as to whether the Americans, for all their material affluence, were really happy.

I began to suspect strongly that, at the very least, what Americans mean when they say they are enjoying life and what we Orientals mean when we speak of happiness are not the same things. To discuss this question in detail would require a volume running to many hundreds of pages, but, to sum it up in a nutshell, the basis not only of American civilization but of Western civilization as a whole is the question of how a collection of people each with a strong sense of individuality can lead a decent social life. As a result, one has the strong impression that they have a constant awareness, from which they cannot escape, of the conflict between the framework imposed by life in society and the desires of the individual.

The method devised to deal with this problem in the West was that in each and every case the individual should give a clearcut yes or no and the final decision be made on a majority basis. In practice, it might well be difficult to find a method that worked better than this. Yet I myself find it constitutionally uncongenial to have to determine my own stand and decide either for or against on every trifling matter. There are, undoubtedly, questions not susceptible of solution unless one does just this. Yet it seems to me that there are many questions where human beings can understand each other and the problem be solved without going to such lengths, without a conflict of views between individuals.

For the Oriental, the self and the world are one. He believes, almost unconsciously, that there is a natural harmony between man and nature.

Western civilization is not the only civilization, and the Western way of thinking is not the only advanced way of thinking. Surely, if the peoples of the world are to lead a truly happy life in the future, a blend of East and West in a new sense will be necessary? This is the kind of question that has preoccupied my mind constantly during the past two years.

2   The Oriental Approach [1948]

I am only too aware of my temerity in giving an address on "Science and the Oriental Approach" to a gathering including so many specialists in Chinese philosophy, literature, and history; I can only ask you to show forbearance for any errors that I may make.

The modern sciences, as needs no pointing out, developed in Europe; there was nothing corresponding to them in the East, and in Japan their rapid development has been a result of importing European science since the closing years of the Tokugawa shogunate. This one fact alone, I feel, largely accounts for the unquestioning assumption of a majority of intellectuals in modern Japan that the Oriental way of thinking is, as a whole, nonscientific and that to cling to an Oriental outlook can only hinder the progress of the natural sciences.

I myself am a Japanese, and as such an Oriental; I was born and bred in the particular historical and geographical setting that this implies, with a particular hereditary disposition, so that however much I may try to assimilate with the West, including the United States and the Soviet Union, there are inevitable limitations. The question of East and West is for me personally an extremely vital one—one, I suspect, that will dog me for the rest of my days. My own experience, however inadequate it may be, leads me to believe that one cannot, as the average intellectual does, simply dismiss the Oriental outlook as inimical to the natural sciences. I myself may, of course, be an exception from which no general principles can be drawn, yet my case may be of some use as one among many.

As for my background and antecedents, my father was closely connected with the Institute of Oriental Culture. He had always been interested in Chinese culture, and was extremely interested in the history and geography of ancient China, together with the archeological aspect. He frequently visited China, and was something of an expert on things Chinese. As for his personality, this had two apparently contradictory aspects. Firstly: he liked old things, whatever form they might take, yet at the same time he also liked trying to answer new questions that nobody had so far tackled. As a result of this combination of qualities, he seems to have made something of a speciality of discovering the new in the old. Although both the humanities and the natural sciences are branches of learning, the relative importance of the old and the new in them is different. The natural sciences are, if anything, more concerned with doing something useful to improve man's lot in the future, yet the particular branch (geology) that my father chose as his speciality has as one of its major goals knowledge of the state of the globe in the distant past, before man was even born. And in geography, history, and archeology, subjects that have such close ties with geology, the interest in the past is still stronger. I myself have, basically, no dislike for old things, yet, engaging in a field of studies such as physics, where the sense of constant change is especially strong—and more so in a new field such as nuclear physics—my interest naturally came to concern itself much more with the future than the past. I feel that the particular interest my father came to show in Chinese culture was bound up to a considerable degree with the fact that it was not only incomparably older than the Japanese, but tended also to revere the old—to be conservative in both the good and the bad senses.

The other outstanding characteristic of my father's nature was his extremely abundant powers of imagination. Thinking about it now, this quality seems to be at variance with the fundamental nature of modern studies. The chief feature of modern studies is, of course, their empiricism. As Professor Sueji Umehara (1893–) has just pointed out, it was not until the development of empirical sciences such as archeology that we could obtain certain knowledge concerning ancient times; in archeology, the same respect is accorded technology and induction as in the natural sciences. If the field of verifiable facts is to be expanded, various techniques are necessary, and that one can discover the relations between ascertained facts is first and foremost due to the process of reasoning based on induction from a large number of examples. This type of empirical spirit was in fact the driving force behind science in Europe from Bacon and Galileo up to the nineteenth century.

The opposite of the pragmatic, the technical, and the inductive is of course the metaphysical, the speculative, and the deductive. Metaphysical speculation was much in vogue in both East and West in ancient times. Preeminent were the Greeks and the Hebrews in the West and the Indians and Chinese in the East—though geographically the Hebrews ought to be classified with the East. In the West, the Greek tradition is carried on to this day. Although via the Renaissance the modern age saw the emergence of pragmatic science as opposed to metaphysical speculation, the tradition of ancient times in no sense disappeared. All that happened was that with the age of Galileo and Newton the speculative and deductive developed close ties with the technical and inductive, so that the tendency of the former to get out of hand was checked. To put it differently, they were experimenters as well as theorists; they invented new mathematical methods such as differential and integral calculus, and they constructed telescopes as well.

In the East on the other hand, and in China in particular, technology was held in persistent contempt, and a situation never developed in which the man of superior intelligence could himself take a hand in experiments. It was the fact that the West underwent modernization in the sense just mentioned that, more than anything else, had a decisive significance in the development of science. However, once science had achieved a high level—in the eighteenth and nineteenth centuries—it began to split up into small specialized fields, and it became increasingly difficult for the individual scientist to undertake everything himself. This tendency is particularly marked in the most advanced fields of study such as physics, where in the twentieth century a clearly defined division between the theorist and the experimenter has begun to appear. Einstein, for example, differed from Newton in being purely a theoretical physicist. To put it differently, the result of progress in any field of study is a growing distance between the fact that is directly demonstrated and the theory that is necessary in order to explain it; the theorist, on his side, is obliged to submerge himself in abstract mathematics and reasoning, and the speculative tendency of ancient times has been revived once more. I myself am studying theoretical physics in the present situation, but, looking back to my childhood, I cannot feel that even then I ever had any particular empirical and technological bent. If anything, my elder brother Kaizuka was far more interested in machines, while I merely happened to be good at mathematics. At first glance it seems a reversal of our proper roles that he should now be a historian while I am a physicist; yet in fact it is not so, for the empirical element plays a larger part today in historical study than in theoretical physics.

To go into this question a little more deeply, the empirical tendency, carried to its logical conclusion, means that one must eliminate everything that cannot be directly demonstrated. It means that one must not consider any quantities that cannot be measured experimentally. In practice, however, the theories of physics today involve certain quantities that cannot be measured. Ever since the seventeenth century, physics has advanced steadily along the path of increasing precision and quantification; this has given rise to a marked trend towards abstraction divorced from direct experience, and the mathematical relationship between quantities expressed in abstract terms has become increasingly important. Even this process of abstraction did not, until the nineteenth century, mean parting company with facts that man could observe for himself; the abstract mathematical expressions were at the same time a faithful representation of phenomena occurring in nature. With physics in the twentieth century, however, no such straightforward correspondence is to be found, and only a very small part of the mathematical relationships that emerge from its highly abstract theories can be demonstrated directly.

Now, the process of abstraction implies arranging things in a mathematical, logical form. Taken still further, it means weeding out and getting rid of the contradictions. To be abstract in mathematics means, ultimately, to reduce everything to contradictions. Here contradiction means to affirm and deny a particular proposition. If such a contradiction emerges, one concludes that there was some mistake in the premises underlying one's reasoning, and one tries to get rid of it. If one goes on repeating this process, one ends up with a system consisting of a large number of propositions that do not contradict each other by starting with a number of given premises or axioms. The view that this is the business of mathematics is what is known as axiomatics. The most important thing here is discovering where the contradictions lie and getting rid of their causes.

In this respect, Orientals, and the Chinese in particular, cannot be said to excel. A thoroughgoing rationalism eludes them. The Japanese, too, are the same in this respect. The Indians, on the other hand, though living in the same Orient, are considerably different; mathematics and logic, in fact, were developed in India from an early date.

Where the Chinese and Japanese would seem to excel, and where they pride themselves in excelling, is in the field of intuition—what is called in Japanese *kan*, a kind of sensibility or alertness. On the average, the Japanese would seem to have good *kan*. This has always, however, tended to be seen as opposed to the spirit of science.

Yet it would be wrong to assume that there is no use for intuition in mathematics and the natural sciences. For though there would be no problem if, as a result of singling out a contradiction, it were possible to reject a particular premise without further ado, cases frequently arise, especially in empirical science outside mathematics, where things are not so simple as this. Let us assume, for example, that we have a theory and that the conclusions that it leads to have so far been in agreement with a large number of empirical facts. Now let us assume that a new fact is discovered that contradicts the conclusions deduced from the theory. So long as one concerns oneself only with the contradiction, one has no alternative but to discard the theory. Thus someone proposes some alternative theory. Let us suppose, next, that this new theory convincingly explains the new fact. But let us suppose also that new contradictions arise between the new theory and the facts that were hitherto explainable by the old theory. What should one do in such a case? All is well, of course, if one can discard both of the theories and successfully explain all the facts with some completely different theory, but in practice new theories are not so easily come by. It is more normal to produce a new theory by finding some means of taking the better parts of two opposing theories and getting rid of the worse. In particular, the development of physics since the beginning of the twentieth century has taken this kind of course. In this kind of case, nothing can be done by logic alone. The only course is to perceive the whole intuitively and see through to what is correct. What is important here, in other words, is not so much to weed out the contradictions as to discover a harmony in the whole. One hears a lot about dialectic nowadays, which has its own logic, distinct from formal logic, and which is supposed to make possible the synthesis of contradictions. I have no particular objection to interpreting logic in such a wide fashion, but the fact remains that in order to synthesize contradictions it is necessary first to survey the whole with intuition. Nor is that all. Even the singling out of contradictions as contradictions can only be possible when the contradictory propositions are grasped intuitively and simultaneously. Even within the field of mathematics, there are some who advocate intuitionism rather than axiomatics. The difference between the two standpoints would seem to reduce itself into a question of how far one recognizes the role of intuition.

Yet to have an intuitive grasp of the whole is still not enough. There is no true creation until there emerges something new that has been overlooked so far. It is here, it seems, that what is called *kan* comes into play.

A question that is relevant here is that of what is known as imagination. Science usually tends to be thought of as the direct opposite of imagination, but only by those who know only one aspect of science. As I have just said, the act of creating something new does not proceed only from the things already given. The scientist himself seeks to add to them, in some form or other, something new. In short, by supplementing what he already has with his imagination, he produces an integrated whole. If he succeeds in the attempt, the contradictions will be resolved. He may not succeed at the first attempt, but if he applies his imagination in all kinds of different ways he will eventually reach a true solution. And this is what actually happens. For us scientists, the power of imagination is an important ingredient.

How able, then, are Orientals in this respect? Generalizations are dangerous, of course, but it would seem that Indians are amply endowed with imagination. But how about the Chinese? My reading of works such as *Hsi Yu Chi*[1] in my childhood, and later of *Chuangtse*, had given me the impression that the Chinese are highly imaginative, but I am told by an expert on Chinese literature that this is not so. With his wide knowledge giving him a larger number of actual examples to hand than I could ever command, he is doubtless correct. The imagination of the Japanese does not, certainly, seem to be especially rich. I myself am far from confident in this respect, though I suspect that I have, by birth, a certain amount of imagination. My father was better at observation than abstract reasoning, and his imagination outstripped both of these in turn. Admittedly, there were many cases where it proved to have been misleading. . . . Looking back over the history of theoretical physics, one might almost, to stretch a point, call it a history of mistakes. Of all the theories thought up by so many scientists, the majority were mistaken and have not survived. Only a very few that are correct live on. To consider only those that have survived gives the impression of very steady progress, but without the many failures behind the few successes knowledge could hardly have made any progress. It is precisely the same as evolution in living creatures: they divided and advanced in many different directions, the majority either ceasing to evolve or becoming extinct, while one single strain evolved into man. In my own personal experience a majority of the ideas I had were stillborn. Only a few out of a whole host grew up into tenable theories.

One further point I would like to make concerns the ways in which the intuition and imagination undergo self-development.

There are various possibilities here, but one of the most outstand- ANALOGY
ing importance is analogy. Analogy is the most concrete of the
ways of applying relationships formed within a certain sphere to
another and different sphere. This is one field in which the Chinese
have excelled since ancient times. The oldest form in which it
appears is the parable. In a large number of cases, the arguments of
the thinkers of old depend upon analogy or parable. A similar
tendency was also to be found, of course, in ancient Greece, yet
the development of a more abstract type of logic at an early stage
is apparent in the system of formal logic perfected by Aristotle.
As I said earlier, I was extremely fond of Chuangtse; his writing
is full of parables and paradoxes, and the greatest attraction of all
was the vast world of the fancy that these opened up for me.

Now, the ideas of Laotse and Chuangtse cannot be fitted into
the mold of formal logic, but this does not necessarily mean that
they are irrational. Theirs is a type of fatalistic naturalism very
much like that to which the scientific view of nature may ultimately lead. Natural science has tried to comprehend all natural
phenomena in terms of cause and effect. In the nineteenth century, as a result of the extraordinary advances made by science,
scientists came to believe that everything was bound by the laws
of cause and effect, and that ideas of free will were no more than
wishful thinking on the part of man. Once this happens, the only
course is to settle into a kind of rationalistic fatalism—a description that applies precisely to the ideas of Laotse, and Chuangtse
also. Laotse's "The wise ruler is without compassion; he sees the
common people as straw dogs" has much in common with this
attitude, and I myself was strongly attracted toward this type of
fatalism during my middle school days.

However, with the further developments in physics in the
twentieth century, and the emergence of quantum mechanics in
particular, the meaning of the term "natural law" began to change
considerably. "Law" no longer necessarily implied causality in

the narrow sense; instead, it was realized that there were some cases where any number of different results were possible from a single cause, and that it was impossible to foretell which of them would be realized. Even in such cases, a law of a kind can still be formulated. Where this happens, however, rationalism does not necessarily mean fatalism. On the contrary, it becomes clear that it is up to science to open up the future by its own efforts. Where this point is concerned, the East is undeniably at a disadvantage. In the old China in particular, there had been an overstrong tendency to revere the past. And yet, if one thinks a little more deeply still, ideas of past and future are all to a greater or lesser extent conceived in terms of human life and destiny; where natural laws in the most fundamental sense are concerned, there is neither past nor future. As a matter of fact, in the theory of relativity, we can imagine a four-dimensional world which comprises both space and time and in which the physical laws hold good. Here, time resolves itself into the fourth dimension, on a par with space, where harmony prevails in an eternal state of rest. Here, the emphasis on the harmony of the whole is apparent rather than contradiction and opposition; and one may sense something close to the Oriental outlook.

I am afraid that what has preceded is not well organized and may well be no more, after all, than the self-slanted ideas of a physicist born in the East.

Notes

1. Part of this book has been translated into English by Arthur Waley under the title *Monkey*.

3 Laotse [1964]

The first book of *Laotse* begins with a passage which I, in my amateur way, would interpret as follows:

> The true way—the natural law—is not the conventional way, the generally accepted order of things. True names—true concepts—are not the conventional names, the generally accepted concepts.

It may be that I favor this interpretation because I am a physicist. Until Galileo and Newton discovered a new "way" in physics in the seventeenth century, Aristotelian physics was the generally accepted concept. When Newtonian mechanics was established and recognized to be the correct way, it went on to become the only generally accepted concept in its turn. Physics in the twentieth century began by transcending the conventional way and discovering a new way. Today, this new way, in the form of the theories of special relativity and quantum mechanics, has become the conventional way. Even such strange concepts as the fourth dimension and the probability amplitude have become almost overfamiliar by now. The time has come to find another, nonconventional, way and other, nonconventional, concepts. Seen in this light, the words that Laotse spoke twenty-three hundred years ago acquire a remarkable freshness.

In fact, though, the passage in question has traditionally been interpreted as follows:

> The kind of way that can be expressed clearly is not the absolute and immutable way. The concept that can be stated clearly is not absolute and immutable.

61

On first consideration, this seems to be precisely the reverse of my interpretation. On further thought, the two are seen, in fact, to be not necessarily contradictory. The latter interpretation is probably the correct one, which would more easily be accepted as obvious to philosophers seeking an eternal truth transcending the developments and transformations of science. And one would hardly be justified in assuming *Laotse* was written for the benefit of physicists. Why, then, should I continue, despite everything, to be so attracted to the work?

I was at middle school when I first got to know *Laotse*. My acquaintance began when I got hold of a single-volume edition of *Laotse, Liehtse,* and *Chuangtse* published in a new series of Chinese classics. At the age of five, before entering primary school, I had begun what is known as *sodoku* of the Chinese classics. My teacher was my grandfather. *Sodoku* involved reading aloud after the teacher passages of Chinese text. For the first few years I had almost no idea of the sense. This *sodoku* continued, I believe, until around the time when I was in first grade at middle school. During those eight or nine years, I studied all kinds of texts, beginning with three of the Four Books—*The Great Learning, The Analects,* and *The Book of Mencius*—*The Canon of Filial Piety* replacing *The Doctrine of the Mean*. As might be expected, it is the works studied during that first period of which I have the most vivid memories. The texts, I still remember, were bound in the traditional style, and *The Great Learning*, for example, began with words that, though I was made to read them without understanding the sense, have to an odd degree remained with me ever since. This choice of works was apparently made according to a policy laid down by my father. *The Doctrine of the Mean*, although an orthodox Confucian classic, was not included probably because my father, who specialized in geology, geography, and history, was not fond of abstract thought and held that it would be profitless, if not actually harmful, to his children.

It was to be expected, perhaps, that with such a background of study, I should be attracted by Laotse and Chuangtse—by the secret pleasure they afforded of reading a book to which my grandfather and father had been unwilling to introduce me. There was the pleasure, too, of coming into contact with a freer way of thinking that transcended the rigid framework into which Confucianism forced human thought and conduct.

Nor was that all. The thinkers of ancient Confucianism—with the exception of Tzese (子思, 483?–402? B.C.), the author of *The Doctrine of the Mean*—were preoccupied with man and his society, and neglected almost entirely the natural world that lay about them. With Laotse and Chuangtse, on the other hand, nature was constantly at the center of their thinking. Man divorced from nature, they argued, could not be happy; and man's power to resist nature was hopelessly puny. This, too, in my middle school days, when I found all dealings with other people irksome, was a major attraction.

From early adulthood onward, I came to feel a strong aversion to the theory of man's impotence and the idea that man should submit voluntarily to nature. I came to attach increasing value to the scientific knowledge that man had wrested from nature by his own positive efforts and to the process of refashioning nature for man's benefit that was an outcome of that knowledge.

However, with the appearance of the atomic bomb, my ideas were obliged to undergo a great change once again. Living in the midst of a scientific civilization, we no longer have a sense of impotence in the face of nature in the raw. On the other hand, we are forced now to worry whether man himself may not go under to scientific civilization, that second nature that man has fashioned himself. Laotse's declaration that "Heaven and Earth are without compassion; they see all things as straw dogs" takes on a new and threatening meaning if one sees "Heaven and Earth" as nature, including the second nature, and "all things" as including man himself.

Today, as ever, man's fate remains unpredictable. But for that very reason, man's efforts on behalf of man are all the more meaningful. Here, one is obliged to think, lies the only purpose in life that makes life worth living and that transcends all questions of failure or success.

4   Chuangtse [1961]

Even before going to primary school, I had studied various Chinese classics. In practice, this means merely that I repeated aloud after my grandfather a version of the Chinese texts converted into Japanese. At first, of course, I had no idea of the sense at all. Yet, oddly enough, I gradually began to understand even without being told.

Most of the works I studied were connected with Confucianism, but, with the exception of historical works such as *The Historical Records*, the Confucian classics held little interest for me. They dealt almost exclusively with moral matters, and I found them somehow patronizing.

Around the time when I first went to middle school, I began to wonder if the Chinese classics might not include other works that were more interesting, with a different way of thinking, and I searched my father's study with that in mind. I hauled out *Laotse* and *Chuangtse* and began reading them, and soon found that *The Book of Chuangtse* in particular was interesting to me. I read it over and over again. I was only a middle school boy, of course, and, looking back on it later, I sometimes wondered whether I had really understood it or not, and what exactly I had found interesting.

Four or five years ago, I was thinking one day about elementary particles when, quite suddenly, I recalled a passage from Chuangtse. Freely translated, the passage in question, which occurs in the last section of the inner part of *Chuangtse*, runs as follows:

> The Emperor of the South was called Shu and the Emperor of the North, Hu. [Both characters mean "very fast," "to run swiftly," and the two characters together in Chinese signify something like "in a flash"]. The Emperor of the Center was known as Hun-t'un ["chaos"].
>
> One time, the emperors of the South and the North visited Hun-t'un's territories, where they met with him. Hun-t'un made them heartily welcome. Shu and Hu conferred together as to how they could show their gratitude. They said, "All men have seven apertures—the eyes, the ears, the mouth, and the nose—whereby they see, hear, eat, and breathe. Yet this Hun-t'un, unlike other men, is quite smooth with no apertures at all. He must find it very awkward. As a sign of our gratitude, therefore, let us try making some holes for him." So each day, they made one fresh hole; and on the seventh day Hun-t'un died.

Why should I have recalled this fable?

I have been doing research on elementary particles for many years, and by now more than thirty different types of elementary particle have been discovered, each of which presents something of a riddle. When this kind of thing happens, one is obliged to go one step ahead and consider what may lie beyond these particles. One wants to get at the most basic form of matter, but it is awkward if there prove to be more than thirty different forms of it; it is more likely that the most basic thing of all has no fixed form and corresponds to none of the elementary particles) we know at present. It may be something that has the possibility of differentiation into all kinds of particles but has not yet done so in fact. Expressed in familiar terminology, it is probably a kind of "chaos." It was while I was thinking on these lines that I recalled the fable of Chuangtse.

I am not the only one, of course, who is occupied with this question of a fundamental theory of elementary particles. Professor Heisenberg in Germany, speculating on what lies beyond elementary particles, has used the term *Urmaterie* ("primordial matter"). Whether one calls it "primordial matter" or "chaos" does not matter, but my ideas and Professor Heisenberg's, while alike in some respects, also have their differences.

Recently, then, I have found a renewed fascination in Chuangtse's fable. I amuse myself by seeing Shu and Hu as something like the elementary particles. So long as they were rushing about freely nothing happened—until, advancing from south and north, they came together on the territory of Hun-tun, or chaos, when an event like the collision of elementary particles occurred. Looked at in this way, which implies a kind of dualism, the chaos of Hun-t'un can be seen as the time and space in which the elementary particles are enfolded. Such an interpretation seems possible to me.

It may not make much sense, of course, to fiddle with the words of men of old in order to make them fit in with modern physics. Chuangtse, who lived some twenty-three hundred years ago, almost certainly knew nothing of the atom. Even so, it is interesting and surprising that he should have had ideas that, in a sense, are very similar to those of people like myself today.

Science developed mostly in Europe. Greek thought, it is often said, served in the broad sense as a basis from which all science was to develop. Professor Schrödinger, who died recently, once wrote that where there was no influence from Greek thought science underwent no development. Historically speaking, this is probably correct. Even in the case of Japan since the Meiji Restoration, the direct influence of Greek thought may have been small, yet indirectly at least it has provided the starting point for her adoption of the science developed in Europe; and in this way we Japanese have inherited the Greek tradition.

Concerning what happened in the past I have nothing further to add. Yet when one considers the future, there is surely no reason why Greek thought should remain the only source for the development of scientific thought. The Orient produced all kinds of systems of thought. India is a good example, and the same is true of China too. The ancient philosophies of China have not given birth to pure science. So far, this may have been true. But one cannot assume that it will remain so in the future as well.

Today, just as in my middle school days, Laotse and Chuangtse are the two thinkers of ancient China for whom I feel the most interest and affection. In some ways Laotse's ideas are, I realize, more profound than those of Chuangtse, but the precise meaning of what Laotse writes is far from easy to grasp. His use of words and phrasing is difficult, and even the commentaries often fail to elucidate the obscurities. What one gets, in the end, is only the framework of his thought. Chuangtse, on the other hand, has all kinds of interesting fables; biting irony is balanced by a grand imagination. Under the surface, there is a profound and consistent philosophy. Simply seen as prose, moreover, the work is incomparable. There are many things in Chuangtse, I feel, that stimulate the reader's mind and make it work better. The fable I quoted earlier was in itself almost certainly written, not about a microcosm, but about the great universe as a whole. Quite obviously, it does not deal with the infinitesimally small particles that form the basis of the natural world, nor with the correspondingly small time and space in which they move. Yet in practice I have the feeling that in it one can discern dimly the microcosm that we have arrived at finally as a result of our studies of physics; one cannot dismiss the parallel as a coincidence. When one looks at things in this way, I feel that one cannot say that Greek thought is the only system of ideas that can serve as a basis for the development of science. The ideas of Laotse and Chuangtse may appear to be essentially alien to Greek thought, yet they constitute a consistent, rationalistic outlook that holds much that is still worthy of respect today as a natural philosophy in its own right.

Where both Confucianism and the mainstream of Greek thought grant significance to man's self-determined, voluntary actions, believing them to offer a valid prospect of realizing the ideals that he cherishes, Laotse and Chuangtse believe that the power of nature is overwhelmingly the greater, and that man, surrounded by forces beyond his control, is simply tossed now one way, now the other. During my middle school days, I found this outlook extreme, yet was attracted to it. From my high school days on, I began to find the idea of man's impotence intolerable, and for a long time I stayed away from the philosophy of Laotse and Chuangtse. Yet all the while I cherished at the back of my mind a suspicion that, however unpalatable it might be for human beings, their ideas harbored an incontravertible truth.

Laotse has a passage that runs as follows:

> Heaven and Earth are without compassion; they see all things as straw dogs. The wise ruler is without compassion; he sees the common people as straw dogs.

The brevity and the air of finality are typical of Laotse. Chuangtse, on the other hand, prefers attractive metaphors such as the following:

> A certain man was afraid of his own shadow and loathed his own footprints. So he started running, thinking to rid himself of

them. But the oftener he raised his feet as he ran, the greater the number of his footprints became; and however fast he ran, still his shadow followed him. Telling himself that he was still not going fast enough, he ran faster and faster without stopping, until finally he exhausted his strength and dropped dead. Foolish man: if he had stayed in the shade, he would have had no shadow; if he had been still, there would have been no footprints.

The outlook expressed here is without doubt fatalistic—a mode of thinking usually described as "Oriental"—but it is far from irrational. Indeed, for us who, with the advance of scientific civilization, find ourselves, ironically enough, increasingly hard pressed by time, the story contains an uncomfortable home truth.

Half my mind revolts against this outlook and half of it is attracted by it, which is precisely why it remains forever in my memory. Books make their appeal in many different ways, but I am particularly fond of the kind of work that creates a world of its own in which, if only for a short time, it succeeds in immersing the reader. *Chuangtse* for me ranks as a typical example of that type of book.

5   The Happy Fish [1966]

People are constantly coming and asking me to write some words for them on the traditional strip of paper used for the purpose, or to do a piece of calligraphy for them to frame. In the former case, I can usually get by with a poem of my own, but with a request for calligraphy—where some suitable short phrase from the classics is usual—I have trouble in finding something suitable. In some cases recently, though, I have been writing the

three Chinese characters that mean, literally, "know," "fish," and
"pleasure." When I do so, I am invariably asked to explain the
meaning. The phrase comes, in fact, from the seventeenth chapter,
"The Autumn Flood," of *The Book of Chuangtse*. The general
meaning of the original passage is as follows:

> One day, Chuangtse was strolling beside the river with Huitse.
> Huitse, a man of erudition, was fond of arguing. They were
> just crossing a bridge when Chuangtse said, "The fish have come
> up to the surface and are swimming about at their leisure. That
> is how fish enjoy themselves." Immediately Huitse countered
> this with: "You are not a fish. How can you tell what a fish en-
> joys?" "You are not me," said Chuangtse. "How do you know
> that I can't tell what a fish enjoys?" "I am not you," said Huitse
> triumphantly. "So of course I cannot tell about you. In the same
> way, you are not a fish. So you cannot tell a fish's feelings. Well—
> is my logic not unanswerable?" "Wait, let us go back to the root
> of the argument," said Chuangtse. "When you asked me how I
> knew what a fish enjoyed, you admitted that you knew already
> whether I knew or not. I knew, on the bridge, that the fish were
> enjoying themselves."

This conversation, which looks rather like a Zen question-and-
answer session, is in fact very different. Zen always carries the
argument to a point beyond the reach of science, but the exchange
between Chuangtse and Huitse can be seen as an indirect com-
ment on the question of rationalism and empiricism in science.
The logic of Huitse's manner of arguing seems to be far better
throughout than Chuangtse's, and the refusal to accept anything
that is neither well-defined nor verifiable such as the fish's enjoy-
ment is, of course, closer to the traditional scientific attitude.
Nevertheless, although I am a scientist myself, I find myself more
in sympathy with what Chuangtse wanted to imply.

Very generally speaking, the ways of thinking of scientists lie somewhere between two extremes—between the outlook that will not believe anything that is not verified, and the outlook that will discount nothing that was not verified not to exist or not to have happened.

If all scientists had clung to either one of these extremes, science as we know it today could hardly have come into being. Even in the nineteenth century, much less in the time of Democritus, there was no direct proof of the existence of the atom. Despite this, the scientists who worked on the assumption that there were atoms achieved a far deeper and broader perception of the natural world than those who sought to understand it without such an idea. The history of science makes it absolutely clear that the attitude that will not accept anything that is not already proven is too stringent.

It is equally clear, on the other hand, that the attitude that refuses to discount anything that cannot be completely denied empirically or logically is too easygoing. In the processes of his thinking or experiments, a scientist must carry out an inevitable task of selection. In other words, he must either discount or forget for the moment, consciously or unconsciously, the majority of all the possibilities he can conceive.

In practice, there is no scientist who clings obstinately to either extreme of outlook; the question is, rather, to which of the two extremes one is closer.

The most puzzling thing for the physicist at the moment is the true nature of the so-called elementary particles. One thing that is certain is that they are far smaller even than the atom, but it seems likely that, viewed more closely, they will themselves prove to have their own structure. In practice, it is next to impossible to distinguish such detail directly by experimental means. If one wanted to take a good look at one elementary particle, one would have to find out what kind of reaction it showed when another elementary particle was brought up very close to it. In practice, however, our experiments can give us knowledge of what happens before and after, but not of what happens at the actual moment of the reaction. In such a state of affairs, physicists tend to lean towards one or the other of the two extremes I have had already described. Some of them take the attitude that scientists should confine their consideration to the situation when the two elementary particles are apart, and that there is no point in speculating on the detailed structure of elementary particles. I myself believe, conversely, that it will be possible by some means or other to obtain a logical grasp of the structure of elementary particles, and I am constantly racking my brains for possible answers. The day will come, I believe, when we shall know the heart of the elementary particle, even though that will not be achieved with the ease with which Chuangtse knew the heart of the fish. To do so, however, we may well have to adopt some odd approach that will shatter accepted ways of thinking. One cannot exclude such a possibility from the outset.

In September, 1965, an international conference on elementary particles was held in Kyoto to commemorate the thirtieth anniversary of the meson theory. It was a small gathering of some thirty scientists. At a dinner held during the course of the conference, I translated the exchange between Chuangtse and Huitse into English for the benefit of the physicists from abroad. They seemed to find it interesting, and it amused me to imagine each of them considering which of the two philosophers, Chuangtse or Huitse, he himself was closer to.

6 The Invisible Mold [1970]

From time to time during the past few years I have referred to the laws of physics as an "invisible mold." We know nowadays that the natural world is made up of a number of different types of elementary particles. Any one particle of a certain type—an electron, for example—differs in not the silightest respect from other particles of the same type. Wherever and whenever they are produced, electrons possess precisely the same mass and electrical charge. This is a manifestation, in its most fundamental form of the rule that prevails in the natural world. Nature comprises within itself an invisible mechanism that can produce the same thing in indefinite numbers, and it is this mechanism that I have recently been referring to metaphorically as the "invisible mold."

The other day, though, as I was rereading *Chuangtse*, I was startled to find a metaphor remarkably similar to mine. The passage in which it occurs, in "The Great Supreme," can be rendered roughly as follows:

> If, just as a metalworker was about to cast something, the metal leapt up and said "I am determined to become the finest sword ever made," the smith would certainly consider he had got hold of some sinister material. In the same way, if a human being were to say, "now that I have been born a human being I am determined to go on living as a human being for ever," the Creator would almost certainly consider him a bad lot. Can we not regard heaven and earth as a great crucible, and the Creator as the metal worker? Whether man lives or dies, whatever happens, is it not enough? Surely all that matters is to sleep peacefully and awake refreshed.

Chuangtse sought to overcome the idea of death by resort to the metaphor of man as something that is fashioned in the invisible mold in the vast spaces of heaven and earth and then, when the time comes, refashioned into something else, so that the distinction between life and death becomes unimportant. I myself had been concerned less with the life and death of human beings than with that of elementary particles. Even so, I could not help wondering why the ideas of Chuangtse so long ago should be so similar to something thought by me today. To me, Chuangtse is a truly marvelous thinker.

7   Motse [1962]

I have already described how I first read *Laotse* and *Chuangtse* in my middle school days. Following that, I tried reading *Motse* together with *Hsuntse* and *Hanfeitse*, but for want of sufficient interest or understanding I never got to the end. In recent years, I again felt an urge to try reading Motse properly. But for a long time, not having the books to hand, I did nothing about it. Of all the philosophers lumped together under the "hundred schools," the book that obviously ought to be read was *The Book of Motse*. This impression was strengthened further when, not long ago, I read a book on the hundred schools by my brother Kaizuka.

The philosophy of Motse (墨子, 480?–390? B.C.) may be summed up in the one phrase *chien-ai* ("dual love"). One loves oneself, and at the same time one loves others. One loves one's relations, and in the same way one loves complete strangers. In this way, there is benefit to both oneself and others. All trouble occurs because of our failure to love each other. This, very roughly, is what is meant by *chien-ai*. Here, then, is a passage that shows how Motse's own reasoning goes:

> It is the same even with the thieves of the world. The thief loves his own room but does not love the rooms of others, so he steals from them for the benefit of his own. The robber loves his own person, but not the person of others, so he is prepared to harm the

persons of others for the benefit of his own. The reason why this happens is that people do not love each other. It is the same with the ministers who encroach on each other's affairs and the princelings who attack each other's states. Each minister loves his own authority and not that of others; thus he hurts others' prestige for the sake of his own. Each princeling loves his own state and not that of others; thus he attacks the states of others for the sake of his own. Here lies the source of all the troubles that plague society. When one considers why such things happen, they are all because men do not love each other.

This idea of *chien-ai* sounds very reasonable to us today. The other man or the other country that one is exhorted to love naturally extends now to include all mankind and all countries. At the same time, *chien-ai* has a slightly different shade of meaning from the type of universal love that is found in Christianity, for example. One loves others at the same time as one loves oneself. One loves other countries at the same time as one loves one's own. In this way, both sides reap the benefit. This is somewhat different from loving others without caring for oneself or of carrying universal love to the point of ignoring all personal gain. While the ethical standard *chien-ai* provides may not reach the highest possible, it is something that is capable of emulation by everybody. No approach could be better suited to mankind today, for whom the only way to survival and prosperity is peaceful coexistence. Motse is said to have lived around the fifth century B.C., but he still astonishes with the modernity of his outlook.

What is so easy for us to understand today was apparently difficult for the ancient Chinese to understand and put into practice. As proof of this, there is the following passage in *The Book of Motse*:

However, the intellectuals of today say, "You are right, of course. To combine the two phases of love is a good thing. But nothing in this world is so difficult as to put it into practice. It is too roundabout." Motse replied, "You may see the advantages, but you do not understand the reasons. To attack fortresses, to do battle, and to lay down one's life for the sake of fame is something that any ordinary man would find difficult. Yet the ruler takes a delight in such things, and his troops carry them out with zeal. How much easier it should be, then, for everybody to love each other and carry out something for the common good. If one asks where is the difference, the man who loves others will almost certainly be loved by others, and the man who benefits others will almost certainly reap benefits from them. If one hates others one will be hated in return, and if one harms others one will be harmed oneself. Seen in this light, *chien-ai* is a natural thing and not in the least difficult. The only thing that makes it difficult is that those responsible for government do not make it the basis of their rule, and persons of rank do not make it the standard for their conduct.

Here again, it is almost uncanny how Motse's argument is even more pertinent to the modern world than to ancient China.

The Motse who advocated *chien-ai* was naturally a pacifist. He was a rationalist, too. For him, and for his disciples, the rulings of contemporary Confucianism contained many irrationalities that they could not stomach. *The Book of Motse*, for example, contains a passage to the following effect:

> If a parent dies, his body is not shut up in the coffin immediately but is left out while they climb onto the roof, peer down into the well, poke in the ratholes and thrust their hands into the wash tub in search of the deceased. If they really believe that the dead man is to be found in such places, they are extraordinarily foolish. If they search knowing that he is not really there, they are guilty of an extraordinary fraud.

Yet this same Motse showed a stronger theistic tendency than either Confucius or Mencius. For him "heaven," for example, has a far stronger anthropomorphic flavor than for Confucians. More surprising still is the fact that Motse and his followers believed in demons and spirits, and they tried to prove their existence. An example is the following passage:

> King Hsüan of Chou had a retainer known as the Lord of Tu killed. As he went to his death, Tu said, "I die for a crime of which I am innocent. If there is consciousness after death, I will come within three years and show the king what ghosts can do." Three years later, the king went hunting with his lords from all over the country. He took several hundred chariots and several thousand men in attendance so that they filled a great stretch of the countryside. At noon, in broad daylight, Tu appeared in a chariot of plain wood drawn by a white horse, wearing a scarlet bow, chased the king and fired at him in his chariot. The arrow struck the king in the back and he fell over in his chariot and died. This happened before the eyes of the many men in attendance on the king, and those who were farther away heard their report of what had happened. There can be no doubt as to the existence of the spirits.

On this score, Confucius, who said, "I do not talk about the supernatural," was far closer to modern man.

In the Chinese classics, there is much talk of "heaven" the "way of heaven," or the "command of heaven." The author of *The Analects*, wrote "Heaven does not say anything, but the four seasons come and go, hundreds of things are born." In *Motse*, this is replaced by expressions such as "I know that Heaven loves all the common people of the world." In *Laotse*, such illusions are swept aside with, "Heaven and Earth are without compassion; they treat all things as straw dogs [made for ceremonial purposes and discarded when they are no longer needed]." From our point of view today, each view has its own element of truth. What always strikes me as most remarkable in rereading the Chinese classics is how the thinkers of China well over two thousand years ago managed to free themselves from primitive obsessions at such an early stage. The emergence from primitive ignorance took place at an early stage in India, among the Jews, and in Greece also, but the Chinese, I feel, were the first of all to attain spiritual adulthood. Confucius in particular shows a maurity characteristic of the middle and late periods of the individual's life, while Laotse seems to penetrate with terrifying insight into the ultimate fate of man the individual and of mankind as a whole. Compared with them, Motse has a more youthful quality, and at the same time is the closest to the thinkers of the West. Chuangtse, too, has a romanticism that is typically youthful, but in itself it is more Oriental than Occidental.

8   Epicurus [1962]

As a child, I used to amuse myself making miniature landscapes. It started when my mother bought me a cardboard box containing a set of pottery—"clay" might be more accurate—a farmhouse, a bridge, a Shinto shrine gateway, and so on, all in miniature. I put some earth in a flat rectangular tray, spread fine-grained white sand and moss on top of it, and made elevations and depressions to represent hills and rivers. Then I arranged the farmhouse and the bridge in suitable places within the landscape. I told myself that I lived in this small world that I had created, and the idea filled me with indescribable happiness. This "world" of mine was pathetically small and offered little variety, but it was quiet and peaceful beyond compare. Small child as I was, I had little knowledge as yet of the larger world of reality, which makes one wonder why, even so, I should have preferred my small, fictional world to the actual world. What was the chief reason that made the child so love his tray landscape—was it heredity, was it instinct, or was it the result of some infantile experience of which he had no memory? I cannot tell; nor, indeed, does it matter.

Something that does matter, though, something that I can do nothing about, is that the love of "box gardens" (as they are called in Japanese) still survives, albeit in different form, in me today, fifty years later. I have only to stand looking out over the compact garden, hemmed in by clay walls, of some ancient temple to feel that indescribable sense of happiness reawakening in my heart. Some ten or fifteen years ago, I was standing in the cloisters of Westminister Abbey, gazing at an inner courtyard, when I heard the choir singing faintly in the distance, and again the same happiness flooded my heart for a while. It was the same when I saw the inner court of the Santa Croce monastery in Florence. The necessary thing to evoke the happiness in each case was that the garden or courtyard should not be too large, and that conditions should be such as to make one forget the outside world. The garden must be surrounded by buildings or walls or trees, and there must be quiet, an absence of the noises of the outside world. There are any number of old temples in Kyoto with gardens that fulfil these conditions; most of them are Zen temples. To stand looking out over such a garden on, say, a Sunday afternoon just before dusk is for me to become once more the child who loved that miniature world of his own creation.

The attraction I felt for Laotse and Chuangtse in my middle-school days was probably not unconnected with this love, as a small child, for "box gardens." And the craving for some world that was closed and self-sufficient a world of eternal quietude—for this, surely, is the common quality—seems to have lain hidden in the recesses of my mind even after that. In time, I took up scientific research. The world of science is an open world: beyond the realm of the known, there always lies another, unknown realm, and I found myself standing perpetually on the boundary between the two. The border is unstable, forever pushed one way or the other by new discoveries, new doubts. The scientist situated on that border must constantly be summoning afresh his own pioneering spirit. Most people, probably, would say that the closed, restricted type of world, the world of eternal quietude, was irrelevant to such a man.

And yet, the fact that the world in which the scientist actually lives is unlimited, forever open, in ceaseless motion, does not mean that it represents his ideal world. Why is it, precisely, that he is always striving to find more universal laws governing the material world, to find more fundamental structural elements? Surely there lurks somewhere in the scientist's mind, even if he does not realize it himself, a desire for that same world of eternal rest? In my own mind, at least, that desire certainly exists, whatever may be true of other scientists. And this desire and the desire to press ahead into the unknown, "open," world have, it seems, been two forces continually driving me from opposite sides.

If one traces the development of modern Western science back to its sources, one comes eventually to the natural philosophy of Greece, where one finds thinkers even more numerous and more varied in their ideas than the "hundred schools" of ancient China. I have always tended to feel more affinity with these ancients than with the modern thinkers of the West. In just the same way I, for one, feel infinitely more at home with the thinkers of ancient China than with the scholars of the Han dynasty and later.

When I think about it more carefully, of course, I realize that most of my knowledge of the thinkers of ancient Greece consists of fragments gleaned from histories of science and other such works, and I wonder just how well-founded this sense of affinity with the Greek philosophers could have been in someone who, though familiar with the Chinese classics since childhood, had no Latin even, much less Greek. Of the many thinkers of ancient

Greece, I had been particularly fascinated by Epicurus, but here too the justification was slight. My knowledge was barely enough to give me doubts about the customary use of the term "epicurean" as a synonym for "hedonist."

Recently, however, I made a belated decision to read what Epicurus actually wrote; fortunately enough, the Iwanami paperback series of classics includes a one-volume collection of letters and other pieces by him. The "letters," of course, are not newsy exchanges between individuals so much as pieces in which Epicurus uses the letter form to set forth the essentials of his own ideas. In a letter to a certain Herodotus, for example, he provides an orderly exposition of nature that comes close to modern atomism.

The contributions made by the natural philosophers of ancient Greece to the long history of science from ancient times to the present day are innumerable, but the most outstanding is indisputably the idea of the atom. The first names that come to mind in connection with it are Leucippus and Democritus. They are said to have been active in the fifth and early fourth centuries B.C., but we have no detailed knowledge of their ideas, nothing but fragments of their views. Epicurus himself lived about a century later than Democritus. In terms of the "hundred schools," he is roughly contemporary with Chuangtse. His view of nature, it seems, almost unquestionably had its roots in the atomism of Leucippus and Democritus. I do not know how far the theory as we know it is a result of original development by Epicurus, but such questions of scientific history apart, Epicurus's letters are full of ideas that are still interesting to us today. His view of nature has its roots in the conviction that the whole universe is nonborn and nondestructible. This is how he reasons:

> Nothing can be born of nothing, for if it were otherwise, anything could be born at will from anything, and seeds, for example, would become quite unnecessary. Again, if a thing's

becoming invisible meant that it had been destroyed and ceased to exist, then all things would long since have become nonexistent, since what they had disintegrated into would not exist. Again, the entire universe has always been and always will be as it is now, since there is nothing for it to change into.

This view of things is, first and foremost, one that distinguishes clearly between existence and nonexistence. What exists continues to exist indefinitely, even though it may change its form. However, the bodies that are conceived of as having existence cannot exist unless they have somewhere to be and to move in. Thus, besides those bodies, there has to be a void. And if existence is going to continue to exist however much matter changes its form, however much it may be taken apart and recombined, it must be composed of atoms that are indivisible and not susceptible to change. Thus one is led inevitably to an atomistic view of nature, with the atom and the void as its two fundamental concepts.

Concerning the movement of atoms, Epicurus thought as follows:

> When atoms move through the void without suffering collision with anything else, they invariably move at uniform velocity. Namely, it is unlikely that something heavy should move faster than something small and light—provided, that is, that nothing collides with the latter.

It was an amazing insight that enabled him to picture the atom as in continual motion, and as continually moving at a constant velocity except for collisions and downward movement due to weight. It is astonishing that all the assumptions of nineteenth-century physicists concerning the motion of molecules in a gas should have been more or less paralleled more than two thousand years previously by the Greeks.

Epicurus recognized nothing apart from atoms and the void. Even the soul he considered to be a kind of physical substance. In that sense, he was an out-and-out materialist. Yet he taught that to think thus was the way to the attainment of peace of mind. It was foolish, he emphasized, for man to be troubled by ideas of birth and death in a world where there was neither any sudden appearance of something hitherto nonexistent nor any sudden disappearance of something formerly existent. The atom might move about unceasingly, yet it remained forever unborn and undying. Epicurus's world is, in short, a kind of world of eternal quietude. In this respect, his ideas resemble those of the philosophers of the East, especially Laotse and Chuangtse. But there is one obvious, basic difference. From first to last, Epicurus insists on the distinction between existence and nonexistence. Laotse and Chuangtse start from the point where the distinction between existence and nonexistence does not yet exist. They believe that "chaos" is still more fundamental than bodies with form. They believe that the way to peace of mind is not to see existence as continuing to exist as such, but to see all existing things as returning, sooner or later, to nonexistence.

The essential correctness of the ancient Greek atomist view of nature has become increasingly clear with the development of modern science. The assertion that the material world in which we live has a permanence that supports the atomist type of outlook still holds good in the light of recent science. Yet by now it is becoming impossible to gain an overall understanding of the physical world so long as one stresses only the aspect of permanence. As man has pushed his explorations on from the world of the atom into the world of the elementary particle, the distinction between existence and nonexistence has become increasingly unclear. An elementary particle that has existed so far may disappear, and different kinds of elementary particles may be born; this kind of transformation is occurring continually. Such is the world of the elemental particles; and when one looks further still, beyond the realm of the various kinds of elemental particles, one finds a world that looks as though it might more aptly be described as "chaos." There too, we shall probably detect an atomistic permanence in some sense or another; but to say this is not the same as refusing from the start to consider anything that does not have an atomistic type of permanence. Rather, we have to discover a certain law of permanence in the midst of the chaotic world. This will give us the reason why things exist.

9 The Tale of Genji [1963]

In the autumn of 1962, Emilio Gino Segre, the eminent physicist, visited us at our institute. At present a professor at the University of California, he was awarded a Nobel Prize for his discovery of the antinucleon. He was born in Italy, and was Fermi's first student when he became professor at the University of Rome. I happened to talk with him of the time, some ten years previously, when I had visited the celebrated Tivoli gardens on the outskirts of Rome, with their waterfalls and fountains and their thronging sightseers. It proved that his father had been a kind of steward or chief retainer to the landlord of the Tivoli; he had been born there himself and had played in the gardens constantly as a child. Yet in telling me of his impressions of Kyoto, he had this to say: "I feel that in every respect things Japanese are more beautifully and delicately made than things Western; on the other hand, everything Western is more practical."

I replied, in effect, that I did not think at all that everything Japanese was more beautiful, but that it was undeniable that from ancient times the Japanese, more than any other people, had never been content that a thing should be *merely* practical. It is a fact that our ancestors sought after beauty, and strove to create beauty, in every detail of their everyday lives, and that the same tradition still lingers on today. It is something that the Japanese should continue to cherish in the future.

The work of literature that comes to mind immediately whenever I consider questions such as these is Murasaki Shikibu's *The Tale of Genji*. It is quite impossible to grasp in all its breadth the extent of the influence exerted on Japanese culture, both directly and indirectly, by this tale during the nearly one thousand years since it was created by its authoress.

Seen even from the viewpoint of world literature as a whole, there is a great deal in *The Tale of Genji* of which Japan can be proud. The novel was written around the turn of the tenth and eleventh centuries. At that time, no other part of the world had produced anything that could be called a "novel." Moreover, whereas a large majority of other cultural products have their origins in China, India, or Western Europe—or farther back still, in Mesopotamia or Egypt—this story is one thing that is genuinely and indisputably Japanese. I have long believed that if I were to select one figure alone from Japanese history to represent the whole, it would be Murasaki Shikibu. According to my taste, I have written at various times of various works from Japan, China, and the West that have created an impression on me, and it has been obvious to me that I could not overlook entirely this central work in the national literature, a work of original genius. Yet every time I turned to the task my mind was seized with reluctance. So many others since ancient times have studied *Genji* exhaustively, and from very many different angles; it seemed unlikely that at this stage one could have anything new to say. All I can do, then, if I am to write about *Genji* at all, is to set down some of my memories in connection with it.

I must have been in the final grades of primary school at the time. My second eldest sister had left girls' school and was studying Japanese and Chinese literature at senior high school. I came home one day to find her setting her notes in order. She had a notebook in which she was making notes with green and brown ink in neat *kana* script. She was copying out the notes she had made at school during lectures on *The Tale of Genji*. I myself had read none of *Genji* at the time and merely had a vague impression of some remote, exclusively feminine world of beauty. I had already by then read quite a lot of the classics of Japanese literature available in handy editions current at the time, but I could not bring myself to tackle *The Tale of Genji*.

Some twenty years passed. Then there came a period, during my thirties, when I was traveling every day from my home, which lay between Kobe and Osaka, and Kyoto University, which meant an hour in the train each way. In the morning, when I was fresh, I sometimes read books connected with my work, but in the evening I was too tired. Thus it occurred to me that I might start reading *The Tale of Genji*. At first, I found it hard-going, as I had expected, but gradually, reading for an hour every day in the swaying train, I began to feel my interest aroused. And oddly enough, as that happened, I found myself being absorbed effortlessly into the world of the story, without worrying too much over the structure of the sentences or over who was talking at a certain point or who performed a certain action. But the world revealed to me was essentially different from any that I had found in books hitherto. It differed from the Western world created by the great works of Tolstoy, Dostoevsky, or Goethe, and it differed equally from the world of the Chinese historical romances and the popular stories of feudal Japan. Just as I had once immersed myself in these latter forms of literature, so now for a while the world of *Genji* took me out of the world of actuality. Indeed, the world about one at that time—as I remember, it was roughly the period from 1941 to 1943—was hardly compatible with good sense or intelligence, and I have no doubt that it was precisely for that reason that I sought to escape from reality in *Genji*.

In this way, I read as far as the "Tamakatsura" chapter. By this point, however, the number of characters had increased considerably and their relationships had become complex. This must have gradually made me feel it a bother to continue reading, or there may have been some other reason, but either way I gave up my reading for two or three days. Once that happened, the thread was suddenly broken, and I did not take it up again until quite recently. People have long referred to "turning back at Suma," meaning that it is common for readers of *Genji* to get no farther than the "Suma" chapter. Since I got farther than that, I was at least rather better than average, it seems, but that hardly qualified me to discuss *Genji* as a whole.

A few years ago, however, *The Tale of Genji* began to appear in a new edition in the "Japanese Classical Literature Series" so I decided to get a copy and read it when my work allowed. Doing so, I found that I had forgotten most of those parts that I had read before. On the other hand, it all came to me as fresh as though I were reading it for the first time. The edition that I used was extremely thoughtful to the modern reader, with the subjects of sentences and necessary passages of explanation between sentences inserted in small type, so that I got a much more concrete grasp of what was going on than on my first reading.

In one chapter, entitled "The Village of Falling Flowers," there occurs the following passage:

> He went without any outriders and took care that there should be nothing to distinguish his coach from that of an ordinary individual. As he was nearing the Middle River he noticed a small house standing amid clumps of trees. There came from it the sound of someone playing the zithern; a well-made instrument, so it seemed, and tuned to the eastern mode. It was being excellently

played. The house was quite near the highway and Genji, alighting for a moment from the carriage, stood near the gate to listen. Peeping inside he saw a great laurel-tree quavering in the wind. It reminded him of that Kamo festival long ago, when the dancers had nodded their garlands of laurel and sun-flower. Something about the place interested him, seemed even to be vaguely familiar. Suddenly he remembered that this was a house which he had once visited a long while before. His heart beat fast. . . . But it had all happened too long ago. He felt shy of announcing himself. All the same, it seemed a pity to pass the house without a word, and for a while he stood hesitating. Just when he was about to drive away, a cuckoo flew by. Somehow its note seemed to be an invitation to him to stay, and turning his chariot he composed the following poem, which he gave into Koremitsu's hands: 'Hark to the cuckoo's song! who could not but revisit the hedge-row of this house where once he sung before?'[1]

It is unlikely that Murasaki Shikibu in this passage was conscious of employing any particular technical devices; she merely inserted within one episode another smaller episode, quite naturally; yet the resulting effect is quite indescribable and transcends mere technique. There is something in common with the appeal of Chekhov's short stories; but nine hundred years separates Chekhov from Murasaki Shikibu.

My reading of the original text of *Genji* twenty years ago stopped, as I have already said, at the "Tamakatsura" chapter. Reading on beyond this during a recent convalescence, I found that the true nature of *Genji* as a full-length novel becomes increasingly apparent from this point on, and that the decriptions become more detailed and richer.

In the "Typhoon" chapter, there is an account of how Yugiri—the son of Hikaru Genji and Aoi no Ue (Lady Aoi)—goes out during a typhoon, at his father's command, to visit various houses and inquire after their welfare:

Towards dawn the wind became somewhat dank and clammy; before long sheets of rain were being swept onward by the hurricane. News came that many of the outbuildings at the New Palace had been blown to the ground. The main structure was so solidly built as to defy any storm. In the quarters inhabited by Genji there was, too, a continual coming and going, which served to mitigate the strain of those alarming hours. But the side wings of the Palace were very sparsely inhabited. Yugiri's own neighbour, for example—the Lady from the Village of Falling Flowers—might easily be by this time in a pitiable state of panic. Clearly it was his duty to give her his support, and he set out for home while it was still dark. The rain was blowing crossways, and no sooner had he seated himself in his litter, than an icy douche poured in through the ventilator and drenched his knees.[2]

Immediately after this, during the typhoon, Yugiri, who is fifteen, meets his stepmother, the beautiful Murasaki-no-Ue, for the first time.

The town wore an inconceivably desolate and stricken air. In his own mind too there was a strange sensation; it was as though there also, just as in the world outside, the wonted landmarks and boundaries had been laid waste by some sudden hurricane. What had happened to him? For a moment he could only remember that it was something distressing, shameful. . . . Why, it was hideous! Yesterday morning. . . . That was it, of course. He was mad; nothing more nor less than a raving lunatic. He had fallen in love with Murasaki!

He did indeed find his neighbour in the eastern wing sadly in need of a little support and encouragement. He managed however to convince her that the worst danger was over, and sending for some of his own carpenters had everything put to rights. He felt that he ought now to greet his father. But in the central hall everything was still locked and barred. He went to the end of the passage and leaning on the balustrade looked out into the Southern Garden.

Even such trees as still stood were heeling over in the wind so that their tops almost touched the ground. Broken branches were scattered in every direction and what once had been flowerbeds were now mere rubbish heaps, strewn with a promiscuous litter of thatch and tiles, with here and there a fragment of trellis-work or the top of a fence. There was now a little pale sunshine, that slanting through a break in the sky gleamed fitfully upon the garden's woebegone face; but sullen clouds packed the horizon, and as Yugiri gazed on the desolate scene his eyes filled with tears.[3]

The fury of the typhoon and even the description of the damage that follow are inextricably interwoven with the psychology of the earnest, warmhearted young Yugiri, and so build up a world of great grace and refinement.

A passage such as this, however, has a great deal of description of natural phenomena and objective facts, which is, moreover, related so precisely that one senses a dissipation of the "Genji mood" compared with other parts of the book. The general impression that one receives from the work is somewhat different. All things, all human beings, move, swaying slowly, through a vague light. Occasionally, a strong light will be thrown on the feelings of one of the characters, and the subtle shifts in his emotions will be thrown into relief. Yet all the while his physical features remain vague. The outlines and textures of things and human bodies are absent. Quite commonly, in other literature, we have a clear conception of a character's outer appearance while the inner recesses of his heart remain obscure; here, the light places and the dark places are reversed. Murasaki Shikibu's discovery, one thousand years ago, was that such a reversal could create a world of unparalleled beauty.

Modern science is unrelenting in its pursuit of accuracy seeking to clarify material existence to the utmost limits possible. The pursuit is all the more unrelenting because of the recognition that beyond the clarity there always lies a new obscurity. In science too, as in most literature, a dim background looms beyond the clearly defined foreground. The world of modern science and the world of *The Tale of Genji* are opposites in their presentation of light and shadow. Whichever may be the positive and which the negative of the picture, it is surely one of the great pleasures of being human and being alive that one can enter into either of these two worlds at will.

Notes

1. Lady Murasaki, *The Tale of Genji*, trans. Arthur Waley, 2 vols. (London: George Allen & Unwin Ltd., 1935), 1:226–27.
2. Ibid., p. 531.
3. Ibid., pp. 531–32.

10   The Freshness of Mellow Ideas [1968]

Looking back on my early childhood, it seems to me that my brain began to absorb the complicated Chinese characters, Chinese words, and Chinese texts at almost the same time as Japanese written in the simple *kana* syllabary. The way in which such things have stayed in my mind suggests that there was already a considerable accumulation at an early stage, probably by around the age of ten. This is somewhat unusual for a Japanese of my generation. It would have been more the rule than otherwise for a samurai before the Meiji Restoration, or a member of a former samurai family thereafter, or for a doctor, or the children of a Confucian scholar, but the experience of myself and my brother was, I suspect, unusual among those born in the latter years of the Meiji period, the early Taishō period, or later.

The memories acquired at an early age come to life again as one gets older. It is difficult to judge the influence that these things had on my life and thinking, but recently I have come to the conclusion that it was more important than I thought in my youth.

Father would tell us from time to time that a child should not be content to be childish, and that we should grow up as soon as possible. This, which was considerably different from the usual parent's advice, seems to have had quite an effect on me. During the same period of infancy and boyhood in which I studied Chinese works of the type just mentioned, I also, of course, read many books and magazines intended for children. Around first grade at primary school, I was much taken by Western fairy tales such as those of the brothers Grimm, and by *Akai Tori* [Red bird], by Miekichi Suzuki (1882–1936) and I grew greatly sentimental. Yet during that same period I read still more adults' books—whatever I could lay hands on—and these, I feel, had a still greater influence on me in later years. This was partly an outcome of my father's advice about being adult, but, more basically, it is related to the nature of the culture and ideas of ancient China. Very generally speaking, the Chinese classics are concerned chiefly with advice for adults and the wisdom of adults. Their ideas belong to maturity rather than to youth, and to old age rather than to maturity. I have always had the feeling, in short, that they are the expression of a settled set of ideas and a culture that in ancient times had already matured and grown old. It was not until my middle school days that I learned English and began to find translated novels interesting and to take over ideas from the literature and thought of the West. Where literature is concerned, though, *The Historical Records*, while a work of history, is also outstanding as a work of literature. It may be because I already felt an affinity for such works that in my primary school days I made the acquaintance of the Yūhōdō books, lighting first of all on the historical romances. With the aid of the readings written in small script by the side of the Chinese characters, I avidly read works such as *Shui Hu Ch'uan* (All Men Are Brothers), *San Kuo Chih* (Romance of the Three Kingdoms), and *Hsi Yu Chi* (Monkey) in Japanese translations made in the Edo period. Even now I can still remember, for example, the majority of the names of the 108 heroes who appear in *Shui Hu Ch'uan*. And my childish mind naturally assumed that this was how a novel should be. Later on, when I read Japanese works such as *Hakkenden* by Bakin (1767–1848) and, inevitably, made comparisons, I realized sadly just how inferior they were.

There are all kinds of problems, however, involved in the way I came into contact with Chinese literature—through translations of the historical romances, and Edo-period translations at that. There is the question, first of all, of just how faithful to the original the stilted Japanese of the translations was. The influence of the illustrations done by Japanese artists was similarly far from negligible. I am sure that the illustrations of Hokusai (1760–1849) for example, created a mood that was not in the Chinese original. Yet in part it was their attraction that induced me to read. In short, my approach to Chinese culture in my childhood was made, as it were, at a remove, in that I was absorbing Chinese Confucianism, Chinese learning, and Chinese literature as it had been taken over by the Japanese of the Edo period, which suggests that my comprehension was very different indeed from that of a Chinese reading the classics and other ancient literary works of his own country.

Another factor that influenced me was that as a child I was made to study calligraphy. The classic calligraphists that I was taught to take as my models were, as I remember, Ou-yang Sun (欧陽詢, 557 641) for *kaisho* and Wang Hsi-chih (王羲之, 307?–365?) for *gyōsho* and *sōsho*. Classical Chinese culture, this meant, came to me via this particular route.

The philosophical element was slight in what I learned from China as a child. Confucianism, of course, deals with ideas, and there are ideas in the historical works too. Yet Confucianism is only one, important system among many systems that existed in ancient times. So numerous, indeed, were these other schools of thought that they were referred to in China as "the hundred schools of thought." My grandfather, however, did not teach me about these, probably because my father had laid down what I was to study. I have already said that *The Doctrine of the Mean*, one of the Four Books, was omitted from my studies. *The Doctrine of the Mean* is a product of abstract speculation, a philosophical work of a kind, and advice and examples to be followed in actual life play little part in it. My father, I suspect, felt that such things were unprofitable, if not actually harmful, to a child.

On the whole, the knowledge of China that I acquired as a child concerned principally ancient China, extending up to Tang or Sung times at the latest; of later dynasties such as Yuang, Ming, and Ching I was taught almost nothing. When I grew up, I got to know something of China after Genghis Khan (1167-1227), but I have never gone beyond the general knowledge level.

My education on China thus had a very one-sided bias and was divorced from actuality. Yet the classical instruction given to children in the West similarly concentrates on very ancient periods —on ancient Greece and Rome, rather than on Greece and Italy in more modern times. The two are similar in being primarily classical and character-forming.

Most boys, when they go on to middle school, go through a period of doubt and mental turmoil. Around that period, my interest began to turn away from China and towards the ideas and literature of the West. I read, for example, Tolstoy's *On Life*, then I went on to his other Russian novels, my interest in things Western gradually increasing as I advanced. It was around the same time that I began to read the works of Soseki Natsume and other Japanese authors. But that was not all. I also, while I was at middle school, learned of the existence of the many other Chinese

philosophers besides Confucius, especially Laotse and Chuangtse;
I read them for myself, and found them, although imperfectly
understood, extremely interesting. From then on, throughout my
years at high school and university and later still, I would go back
to them from time to time and reread them.

As I have already said, Chinese culture is extremely ancient, and
I have long had the feeling that the ideas that came into being in
what might be called the "classical ancient period" were no
longer the ideas of youth, nor even of maturity, but of old age.
The philosophy of Laotse in particular is the most aged of them
all. During human society's—or perhaps one should expand it and
say the human race's—long history, all kinds of civilizations have
arisen, only to decline again, in many different areas of the earth,
and I cannot avoid the feeling that Laotse, twenty-odd centuries
ago, had already foreseen the state of human civilization today,
or even the state it will reach in the future. Or perhaps it would
be more correct to say that already, at that time, he had discovered a state of affairs that, though superficially very different,
in fact resembled the situation facing mankind today. Possibly it
was for this very reason that that odd work *The Book of Laotse*,
was written. Either way, it is astonishing how Laotse, living in
an age before the development of a scientific civilization, should
have set forth such a scathing indictment of scientific culture from
the beginning of the modern age onwards.

A similar kind of outlook is to be found in Chuangtse also.
Either way, we today have abandoned the optimistic nineteenth-
century view that scientific progress necessarily promotes the
happiness of mankind, and have come to cherish very fundamental doubts as to where scientific progress is leading. The same
doubts as to the outcome of man's actions were already manifest
in ancient China.

This does not mean that in my primary and middle school days
I was already attempting to peer into the world's future. What is

true, though, is that from that time on I had the feeling that society was a nuisance. I was constantly beset with the feeling that to live among and have dealings with a large number of human beings was a troublesome business. Laotse, I imagine, would have put it in this way: it is not wise for ever larger numbers of people to engage in ever livelier converse, and for men ever farther removed from each other to visit each other ever more frequently. It is better for states to be small, and men few. Even though one hears the cocks crowing and the dogs barking in the next village, it is better to remain aloof.

In practice, of course, life in Laotse's time was extraordinarily different from life today. Human dwellings were scattered and the population was small. What would Laotse and Chuangtse think, then, if they were to see the world as it is now? They might well be stunned by the extraordinary complexity of human relations brought about by the development of the mass communications media and the transport networks. This is particularly true with regard to what has happened since the occupation. Contacts with foreign scholars have increased, and the number of international conferences has gone up rapidly. Attendance at such conferences goes up steadily, too—from a hundred to three hundred, from three hundred to five hundred, from five hundred to a thousand. The bother of meeting people has attained an international scale. It is an intolerable situation. Once communications have expanded and become more convenient, it is extremely difficult to change the direction of their development and get back to the quiet world of old—a fact the realization of which only aggravates our sense of irritation still further. The ideas of Laotse and Chuangtse represent a clear attempt to get away from this frustrating state of affairs. It was this attempt that gave rise to ideas of hermits and of Utopia, and, in its turn, to such exalted art as the ink landscape paintings. A favorite theme of Chinese ink landscapes is a waterfall deep among the mountains with a hermit's dwelling standing close by and the hermit shown alone, gazing at the waterfall.

The Chinese-style landscape was imported into Japan, and both ink landscapes and the "Southern School" of painting took root in Japan, yet it seems that paintings by those Chinese artists who have been most highly esteemed in China remained unknown, or at least underestimated, in Japan. Among them, I suspect, should be included the Four Great Masters of late Yuan as well as the Four Wangs, Wu Li, and Yün Shou-ping of early Ching. Underlying their work there is a very characteristic Taoist outlook not to be found in any other cultural zone. This type of art, I feel, is one of the finest products of the whole flow of Chinese culture from ancient times on. One wonders why Japanese culture, though subjected to such an overwhelming influence from China should have taken over so little that was associated with the philosophies of Laotse and Chuangtse. It is a question to which I have still not found a complete answer.

Buddhism, which had come, via China, from far-off India, was taken over by Japan and had far-reaching effects on Japanese culture. Japan also took over Confucianism, the teachings of Chu Hsi (朱子, 1130–1200) in particular gaining widespread acceptance during the Edo period. Why, then, did not Taoism, which had had such a following in China itself, take hold in Japan? If one looks hard, one can, of course, detect certain influences, but nothing comparable with those of Buddhism and Confucianism. It is an interesting question.

In his *Sangyō Shiiki*, Kōbō Daishi (774–835) made a comparative evaluation of Confucianism, Buddhism, and Taoism, laying great stress on the superiority of Buddhism. It is interesting that the young priest Kūkai (as Kōbō was then known), still in his early twenties, should have taken up Taoism as an object for comparison with the others. Japanese intellectuals of succeeding ages also seem to have read works such as *The Book of Chuangtse* a good deal; Michizane Sugawara (845-903), for one, turned the *Hsiao Yao Yu* of Chuangtse into his own very distinctive type of verse; yet still Taoism failed to find much general acceptance. Taoism unlike Buddhism and Confucianism, underwent no philosophical development following the deaths of its founders, Laotse and Chuangtse; on the contrary, it tended to degenerate and to become associated with local religious beliefs. It was this religious form of Taoism with which the Japanese came into contact; they had little chance for contact with the ideas of Laotse and Chuangtse, and in this may be what led them to the conclusion that Buddhism was superior. Even so, it is unlikely that a thinker of the level of Kōbō Daishi should have been ignorant of the essential ideas of Laotse and Chuangtse. It seems more likely that, being a young man at the time, he was dissatisfied by their negativity. Even in Buddhism, it was in the most positive of its forms—Esoteric Buddhism—that he was to choose to expound himself.

Another factor not unrelated to what I have just written is that even today the Japanese, are, I feel, a young nation. This is not to indulge in value judgements concerning spiritual ages; quite apart from such evaluations, it seems likely to me that, just as in individuals, one can distinguish between periods of youth, maturity, and old age where patterns of thought and the human spirit are concerned. Japan as a country is quite old, yet still, I feel, shows a marked youthfulness. She is youthful—if not childlike—in the extraordinary curiosity she has always shown towards the things of other countries. Today, I suspect, China herself has undergone a rejuvenation, yet the ancient China that produced thinkers such as Laotse was, surely, already in its old age. May it not be precisely because the ideas that it produced permitted it to see far into the future that those ideas, unlike those of ancient Greece, did not give birth in turn to a scientific civilization? At one stage in my life, I myself took leave of the world of Laotse and Chuangtse and turned to the world of physics, but since I reached middle age their ideas have, unmistakably, acquired a new life in my mind.

Ancient China, in short, finds a place inside me in all kind of ways. While that implies contradictions with the fact that I am a scientist, it also serves, conversely, to give me, as a scientist, some individuality. I cherish within me, of course, not only China, but a very great number of traditional Japanese things and not a few Western things as well. All of them have served to enrich my life. But since my sixtieth birthday it is the aged, mellow ideas of ancient China that I have felt most affinity for. At the same time, those ideas also seem to me now to be extraordinarily modern.

Nobody finds the world arranged to suit his own convenience in every respect. Neither for human beings nor for other living creatures can the world be said to exist on their particular behalf. Nor is one born into this world of one's own volition: one is oneself before one knows what is happening. And whether one finds the world to one's liking or not, one cannot live forever. Nor, though, can the time of one's death be predicted in any but very extreme cases. In this respect, one is in the same boat as all one's fellows.

But it is precisely because we have been born into such a world that we should give comfort and aid to each other. If that is not possible, then we should at least try not to cause others inconvenience. Nor should we, as far as possible, treat other living creatures cruelly.

This way of thinking has gradually established itself more and more firmly in my mind. Perhaps it is my own mellow brand of a philosophy of old age?

III ON CREATIVITY
AND ORIGINALITY

I   Intuition and Abstraction
     in Scientific Thinking [1964]

My aspiration to be in Greece has finally been realized after so many years and in a quite unexpected way. I never dreamed that I would have the opportunity to give a public lecture on the Hill of the Pnyx in Athens. As a man who has become a physicist, I am very well aware of my indebtedness to the Greeks who initiated the great undertaking to reveal the truth hidden deep in nature. This I certainly share in common with the other speakers. There is, however, one extra thing in my case.

I was born, brought up, and educated in a country which was so far away from the Western world that the influence of Greek philosophy, science, and art was hardly felt until about a hundred years ago, when our ancestors began their wholehearted endeavor to learn science and technology from the Western world. This had an overwhelming effect on the intellectuals in our country, so that most of those of my generation have already been separated to a great extent from the classical Oriental teaching which our ancestors inherited from preceding generations. However, when I was only about five years of age, my grandfather and my father began to teach me, contrary to the fashion of the time, Chinese classics such as the books compiled by the disciples of Confucius. At the age of thirteen or fourteen, I found books on Taoism by Laotse and Chuangtse in my father's library and I was deeply impressed by their philosophy of nature and life which is about as old as the philosophy of ancient Greece.

With such a background, I entered the world of modern science which was based on the heritage of the ancient Greeks. Since then, one question has haunted me from time to time, and it still haunts me. The question is: "Why has science reached its present level as a result of its foundations in Greece and not in any other land?" I cannot help asking myself this question, particularly because the philosophy of Laotse and Chuangtse on nature and life was not only very profound, but also rational and human. They clearly recognized the universal way or natural law prevailing everywhere in nature.[1] Why were they or their followers not able to develop their ideas to a definite shape comparable with that achieved in natural philosophy in ancient Greece? As yet I do not have any very clear answer, but I am sure that it must be closely related to man's ability for abstraction. For some reason or other, China could not produce the type of genius as exemplified by Pythagoras and Democritus. It was undoubtedly crucial for the creation of the prototype of an exact science like physics that there should appear a Pythagoras who grasped natural laws as simple and definite relations between numbers, and a Democritus who developed the idea of the existence of invisibly small atoms and the abstract concept of void.

Of course, this capability for abstract thought, which was not only crucial at the beginning but also continued to be essential for the further development of physics, could not work by itself. It always presupposed the presence of the power of intuition, with which both Greek and Chinese geniuses in ancient times were richly endowed. The important point would appear to be the balance or cooperation between intuition and abstraction. Here lies a problem of our age of scientific civilization—there seems to exist a general feeling of the estrangement of science from other cultural activities such as philosophy and literature. It may sound very strange, but I have felt more and more strongly my own estrangement from the trend of contemporary physics in spite of the fact that I am myself a physicist.

If we consider the history of physics, we recognize two great revolutions in the last three or four hundred years. The first one was, of course, the revolution in the seventeenth century initiated by Galileo and completed by Newton. The other began towards the end of the nineteenth century with some great events—the discoveries of X-rays, radioactivity, and the electron. This second revolution had two peaks: one at the start of the twentieth century culminating in Planck's and Bohr's quantum theory and Einstein's theory of relativity, and the second when quantum mechanics was established in the nineteen-twenties. As to if and when the third revolution will begin, the opinions of physicists may differ very widely. I shall try to give my own answer later, but for the moment I would like to draw attention to the following remarkable effect of the second revolution.

This is the wide departure of theoretical concepts in physics from intuition and common sense. In other words, a trend of abstraction has become conspicuous during the course of the development of physics from the beginning of the twentieth century. Physicists are compelled to accept abstract mathematical concepts when they are logically consistent and their consequences agree with experiments, even if they contradict our intuitive picture of the world.

In this respect, I propose to give only two outstanding examples; according to our intuition, space and time are entirely different from and independent of each other. Thus, we could accept Newton's concepts of absolute space, absolute time, and absolute motion without much resistance in ourselves. Einstein's relativity of space, time, and motion could only be accepted at the cost of the intuition on which physicists had been relying until the beginning of the twentieth century.

However, as time goes on, what seemed initially to be abstract has gradually become something concrete to many physicists and a new sort of intuition took shape in their minds. Nowadays, a four-dimensional space-time world is intuitively grasped by a physicist almost as clearly as Newton's space and time were grasped by those in his time. The new intuition, which was the outcome of abstraction, becomes in its turn a new starting point for further abstraction.

The other example is the evolution of the concept of matter. There were two mutually opposing concepts of matter throughout the history of physics. One was the atomistic concept of matter begun by Leucippus and Democritus; the other was the concept of the material world attributed to Aristotle, who rejected void. Throughout the modern development of physics, the Aristotelian concept of omnipresent ether had prevailed among physicists until it was renounced by Einstein. On the other hand, the atomistic concept of matter was not directly verified empirically until the end of the nineteenth century, when the great discoveries came one after another and reestablished atomism on a solid basis.

Physics in the twentieth century appears to be a succession of triumphs for atomism, but this is a one-sided view of the whole truth. Physicists have long been distressed by the undeniable duality, that is, the wave-corpuscle duality of light. According to intuitive thought, a wave is something extending continuously in space and a corpuscle is something occupying a small, limited region. Light can either be a wave or a corpuscle, but cannot be both. The dilemma became even more acute when the duality was discovered for the electron, which is a constituent of matter.

The solution of the dilemma came with the establishment of quantum mechanics, and our old intuition had once again to be abandoned. Matter, like light, cannot be grasped by application of our old image of corpuscle or wave. We need abstraction, and indeed we have to go much further in this case of relativity. Although in the beginning the mathematical expressions of relativity appeared to be rather abstract, the new intuitive picture of a space-time world has gradually settled down in the mind of the physicist.

Of course, it turned out to be extremely difficult to construct a new image of matter and light which is both intuitive and a faithful representation of what is described by abstract mathematical symbols in quantum mechanics. Eddington, eminent British astronomer and physicist, once used a new word "wavicle" for an object with wave-corpuscle duality, but this did not help much. The inadequacy of the intuitive concepts such as corpuscle and wave led the physicist to have recourse to the notion of probability. Now probability is a concept which is not at all new to physicists, but it used to be introduced only when it was thought that knowledge of the phenomena in question was incomplete because of excessive complication. It was surprising to find that the notion of probability was needed even when such a simple thing as an electron was being considered.

According to the usual interpretation of quantum mechanics, the future position of an electron cannot be predicted uniquely however much we know of its present state. Instead, we are compelled to be satisfied with calculating the probability of its being at a certain position at a certain instant in the future. This clearly indicates the failure of the naïve intuition of a corpuscle when applied to the electron. On the other hand, the intuitive picture of a wave as applied to the electron also fails, because the electron keeps its individuality and remains an entity, whilst a wave will eventually spread over the whole of space. The intro-

INTUITION AND ABSTRACTION · 105

duction of the concept of probability, nevertheless, does not imply the complete abandonment of the old intuition.

However, the real situation is much more serious since abstraction in quantum mechanics went much further than the mere introduction of the concept of probability. The primary feature of quantum mechanics is not probability itself, but the even more abstract notion of probability amplitude, which is generally expressed by a complex number. From here the return to intuition may seem an anachronism.

Thus, in discussing two examples, I have given a brief sketch of the course taken by theoretical thinking during the first thirty or forty years of this century. Now, another thirty years have elapsed. If we look superficially at the development of physics during this period, it is again full of new discoveries and great achievements. For example, a host of new particles has been discovered—of these, only a few, such as the positron, π-meson, neutrino, and antiproton,[2] had been anticipated by theoretical physicists in the previous period. All the other particles were successively discovered in the course of a few years around 1950, mostly in cosmic rays, although no one had anticipated their existence by theoretical reasoning. During these few years, many particles of extremely short lifetime were discovered using big accelerators, and physicists have not yet been able to give a profound and convincing reason for their existence.

Thus, in contrast with the brilliance of scientific thinking in the early years of our century, which was capable of making predictions, most of the physicists of today are at a loss to know what to do with the richness and complexity of the newly explored world of subnuclear physics. It seems as if present-day physicists have lost the gift of foresight inherited from their forerunners.

This rather disappointing change in the outlook of physics seems to me to be very intimately related to the theme of intuition and abstraction in scientific thinking. In order to elucidate the point, let us once more consider a concrete example. The greatest figure in this field during the early years of our century was Einstein. His theory of relativity abandoned the old-fashioned intuition of absolute space, absolute time, and omnipresent ether, so that Michelson's experiment on the velocity of light would not be contradicted. However, he was not simply motivated by, and satisfied with, the abstractions which led to logical consistency and agreement with experiment. He aspired to a new kind of beauty and simplicity in nature which was yet to be discovered—abstraction is always a means of simplification, and in some cases a new sort of beauty appears as the result of simplification. Einstein had a sense of beauty which is given to only a few theoretical physicists. It is very hard to tell what a sense of beauty means to a physicist, but it can at least be said that simplicity alone may be reached by mere abstraction, whilst a sense of beauty seems to guide a physicist in the midst of abstract symbols. The geniuses of ancient Greece were endowed with just that sense of beauty which theoretical physicists need today—it is closely related to man's marvellous ability for pattern recognition, which cannot easily be imitated by a machine such as an electronic computer. I always wonder how, in the midst of the crowd, one can easily and almost instantaneously recognize somebody with whom one is acquainted. It is destined to be an extremely complicated procedure if it is replaced by the composition of a great number of dichotomic or yes-no processes, which are used in digital computers.

In 1905 Einstein established a special theory of relativity which all physicists have come to accept because a great many empirical facts supported it. Ten years later he succeeded in constructing a general theory of relativity which was also amazingly confirmed within a few years by astronomical observations. However, the number of empirical facts which were decisively for general relativity and against other possible theories remained very small for

many years. Under such circumstances it is still possible for physicists to withhold wholehearted support from general relativity,
and to pursue some other route in order to reach agreement with
the crucial empirical facts. In fact, many of the younger physicists
are reluctant to accept the way of thinking which led Einstein to
the establishment of general relativity. Here we find a very interesting example of abstraction versus intuition which is relevant
to the principle of simplicity and the sense of beauty mentioned
above. I would like to discuss this point in some detail.

I have already said that the abstract mathematical concept of
quantum mechanics could not be adequately replaced by an intuitive picture, although the latter could help in understanding
the world of atoms provided that we already know the abstract
concept itself. This trend of abstraction has gone to extremes during the last twenty years. The probability amplitude, which is a
complex number and which does not correspond directly with
any intuitive image, has become almighty or something absolute
to most theoretical physicists today.

Thus, to some physicists of the younger generation, theoretical
physics is reduced to the mathematics of complex functions of
complex variables supplemented by mathematics of abstract
groups. I do not deny that further and further abstraction on our
part may be necessary when we go deeper and deeper in pursuit
of nature. Nevertheless, I feel very uneasy about the fact that this
one-sided trend to abstraction lacks something which is very important to creative thinking. However far we go away from the
world of daily life, abstraction cannot work by itself, but has to be
accompanied by intuition or imagination. Einstein had a lofty
imagination about our space-time world which enabled him to
construct an amazing theory of general relativity and of gravitation. It was not only simple and beautiful, but had the power to
foretell.

In contrast to it, the theoretical concepts of contemporary subnuclear physics are much more abstract and depend upon very advanced mathematics, yet we feel nothing sublime in them comparable with that felt in the general theory of relativity. This is intimately related to the fact that in the last twenty or thirty years there has been no real revolution in the way of thinking of physicists. In other words, the fundamental concepts of the special theory of relativity which were established in 1905, and those of quantum mechanics which followed in the twenties, were not in essence altered, but only the abstraction progressed. The power to predict is almost lost and theoretical physics has largely been reduced to a general and abstract way of describing what is already known empirically. During the sixty years of this century, theoretical physics has become less and less romantic, so that we may even think that we are now in an antiromantic era. Is it inevitable that a branch of science should become irreversibly older?

The history of science fortunately provides a number of examples which show that this is not inevitable. At the beginning of our century, physics was capable of rejuvenation. Another rejuvenation of fundamental physics may be expected if greater regard is given to intuition or bold imagination as a supplement to the inevitable trend to abstraction.

Speaking in this way, I cannot help tracing the history of science back to ancient Greece. There, not only were intuition and abstraction in complete harmony and balance with each other, but there was also no such thing as the estrangement of science from philosophy, literature, and the arts. All these cultural activities were close to the mind and heart of human beings. A person could appreciate poetry in the same vein as he appreciated geometry. To a scientist like myself, natural philosophy in ancient Greece looks as romantic as the stories of gods and goddesses, and the tragedies of heroes. In sharp contrast to this, physicists today are apt to be overwhelmed by huge accelerators and by complicated high-speed electronic computers. It seems as if one of the prime aims of fundamental physics is to obtain a great number of data from a big accelerator and then to put them in a high-speed computer to analyze and compare the results with theoretical formulae. However, I do not think that the pursuit of truth in nature will come to an end in this way. I firmly believe that the bold imagination of men is at least as important as the big machines in the effort to reveal the underlying truth in nature. In this connection, I find it very encouraging that biological science is making rapid strides in the deeper understanding of life. There, the creative activity of each scientist can play a decisive role. In contradistinction to the world of subnuclear physics, biological science is, as yet, far from being overridden by big machines.

Plato explained learning, or the acquisition of knowledge in this life, as a process of recollection. To Plato, it was the recollection of the life which a man had experienced before he was born into this world. To us, the scientists in the middle of the twentieth century, it could be the recollection of ancient Greece, and also, in my own case, the recollection of ancient China. Such a recollection on such an occasion as this might give us a new inspiration for the deeper understanding of nature.

Notes

1. The word in Chinese corresponding to "nature" means literally "that which is so by itself." In the book *Chuangtse*, the following sentences are found: "The sage attains *Logos* inherent in all things on the basis of beauty in Heaven and Earth." When Chuangtse was asked to answer the question "Where is the Way?", his reply was "Everywhere." He was further asked to give examples and he answered "The Way is in insects, plants, tiles and walls," to the great surprise of the questioner.

2. Dirac had arrived at the new concept of the antiparticle before the positron and antiproton were discovered. They turned out to be the antiparticles of the electron and proton respectively. The neutrino is the particle whose existence had been postulated by Pauli in order to account for the beta decay of the radioactive nucleus, long before it was confirmed by experiment. The π-meson turned out to be the particle whose existence had been deduced from theoretical considerations of the nature of nuclear forces.

2   Creative Thinking in Science [1967]

Man and his world is a theme which has many and immensely diverse aspects. Nobody can take up this theme in all its aspects. In this lecture I will confine myself to man and his science. Even this is too broad to be dealt with in a limited time; I shall further narrow the subject later on, but for a while let us dwell on the role played by science in shaping the world as the environment of man.

What we call environment today is not simply the part of nature which happens to surround man. Man builds houses, manufactures clothing, and creates machines. Science and technology intervene between man and nature. Today man is in the midst of man-made articles, mixed up with nature in its original form. Man calls the former "artificial" and the latter "natural," but we should not forget that they are not really distinct from each other. On the contrary, they are essentially the same for the following two reasons: firstly, the material is the same—the raw material of any man-made article had been there already in nature before man appeared and touched it. Secondly, the same laws of nature prevail on all things irrespective of whether they are "natural" or "artificial." One may well say "man created science," but one must not make this mistake. Man could not create science from nothing. What man could do in creating science was to discover something hidden in nature. The two most important things which man has to discover in nature are the raw material in its most fundamental sense and the universal laws of nature.

Now the question arises, "How can man ever discover them?" We know that scientists themselves have given answers to this question. Perhaps the best-known answer is that of Galileo Galilei, who was one of the great initiators of modern science. He said that experience and reason are the two pillars on which science is built. This is very true, but how can experience and reason work together in achieving new discoveries in science? Each one of us is accumulating experiences day by day. We are making efforts to think according to reason as much as we can, at least when we are engaged in scientific investigations. However, it is only on a very rare occasion that a scientist discovers something of importance. Galileo taught us that the mere acceptance of stimuli which come from the environment and which result in the accumulation of memory in the brain is insufficient for our purpose. He taught us by his own achievement that experiments purposely designed are of vital importance in uncovering the truth hidden in nature. He asked questions addressed to nature and succeeded in receiving answers directly from nature. In doing so, he only needed rather simple devices.

Today physicists are carrying out experiments in the same spirit as Galileo's, but one should not and, in fact, cannot overlook the immense change in the way of doing important experiments which has occurred during the past twenty years or so. Let us take for example the big accelerators which are the devices used to accelerate particles such as protons or electrons up to a speed very close to the velocity of light—three hundred thousand kilometers per second. These protons and electrons together with the neutrons are the common constituents of ordinary matter. They are collectively called the elementary particles. In a sense they are the only constituents of matter, being the raw material in its most fundamental sense.

In another sense, however, this is an overstatement, for the following reason. First of all, matter and energy are no longer entities which are completely separated from each other; rather, matter can change into energy and vice versa. According to Einstein's theory of relativity, matter is to be regarded as a form of energy. Furthermore, according to Planck's quantum theory, light, which is a form of energy, consists of small units of energy, or energy quanta. These energy quanta behave like very small corpuscles, so that today they are called photons. Thus the photons are to be included in the family of elementary particles, but even they are not the last members of the family. There are many other forms of energy which are distinct from ordinary matter as well as from light or electromagnetic energy in general.

Accordingly, physicists have come to think of the possible existence of such particles as neutrinos and mesons, and have confirmed their conception by experiments. Here we see a very interesting new feature of experiments concerned with elementary particles. A typical example is the case of mesons. They can be found experimentally because they have been created somehow either naturally or artificially. Mesons do not exist alone in ordinary matter. They are created naturally in cosmic rays, but they decay eventually into electrons and neutrinos.

In order to create mesons artificially, it was necessary to construct an accelerator about ten times larger than the largest one which existed before the war. This was the beginning of an almost endless race of constructing larger and larger accelerators in the twenty years following the war. The race is still going on. What was the fruit of this very expensive race? One may well say the fruit was rich enough. Physicists created and discovered a great variety of new elementary particles, most of them unexpected. I have already said that physicists cannot create something from nothing. They could create new particles by making use of accelerators, because there was a reason on the part of nature to change the more familiar form of energy into a form which was new to man. So one may again say, "Man discovered new things hidden in nature." What we physicists do not know yet is the very reason why nature acts in such a complicated and unexpected manner, giving rise to a great variety of new particles in response to man's questioning by way of experiments. We have to think very hard in order to find out the reason; which means, in short, we have to find out the reasonable and unified description of all kinds of elementary particles.

Perhaps I have been dwelling a little too long on the present state of affairs in fundamental physics. I dwelt on it because it is relevant to the theme of my talk. However, I must now begin to deal with the subject (creative thinking in science) which may be of interest not only to physicists but also to those who are not concerned specifically with the frontier of physics.

This subject is a way of thinking in science which can be creative in the sense that it results in the discovery of a truth in nature hitherto unknown to man. We have been taught that there are two methods of thinking in science. One is the method of induction which was advocated by Francis Bacon and effectively used by Galileo Galilei. It starts with the comparison of more or less similar experiences or experimental results. The other is the method of deduction which was consciously employed by René Descartes in guiding his own mental activities. It begins with a few self-evident facts or principles. However, it is not at all easy to see where the seed is in each of these two methods which enables man's thinking to be really creative, although we are very well acquainted with these methods.

I have been interested in this problem for many years because I had been frustrated very often in my work in theoretical physics. I have come upon fruitful new ideas only very rarely during my nearly forty-year career. It was more than twenty years ago that I asked a senior colleague of mine, who was professor of psychology of the university in Kyoto, the answer to the problem just mentioned. He suggested the importance of analogical thinking. This was not at all new to me because I was well aware of it already at that time. As a matter of fact, the articles and books dealing with the problem of creative thinking, so far as I know, always begin with the importance of analogy. If we look back in history, we meet with a number of great thinkers or philosophers who appeared more than two thousand years ago in such regions as Greece, Israel, India, and China. They taught people by making ample use of analogy or metaphor. It seems to me that they used it not only for persuading others, but also for finding out truths hitherto unknown to themselves.

The essence of analogy as a form of creative thinking can be briefly put. Suppose that there is something which a person cannot understand. He happens to notice the similarity of this something to some other thing which he understands quite well. By comparing them he may come to understand the thing which he could not understand up to that moment. If his understanding turns out to be appropriate and nobody else has ever come to such an understanding, he can claim that his thinking was really creative.

In the course of development of natural philosophy, which grew up to modern physics, we can find many examples showing the effectiveness of analogical thinking. When Leucippus and Democritus conceived of atoms for the first time in history they must have imagined them as something similar to a visible body such as solid balls, although they are too small to be seen by the naked eye. Newton, in the seventeenth century, seemed to have believed in the existence of atoms as conceived by ancient Greeks, judging from the queries at the end of his book *Opticks*. At the same time, though, he seems to have refrained from constructing his theory of motion explicitly on the basis of an atomistic concept because he thought the latter to be too hypothetical. In the nineteenth century a number of great scientists, such as Dalton and Boltzmann, took advantage of the presupposed similarity between the invisible atom and the solid body which is familiar to man. In the early years of the twentieth century the structure of the individual atom itself had become a matter of concern, and different atom models were proposed by physicists such as J. J. Thomson and Nagaoka. The way of thinking which makes great use of models is rather common in physics and chemistry, but this is just a typical example of analogical thinking. When Rutherford proposed a better-founded atom model than others, he again made use of the analogy with the solar system. However, it turned out that his answer to the problem of the structure of atoms was not the final one.

At this stage an essentially new feature was to be added to analogical thinking, that is, the recognition of dissimilarity which lies side by side with the similarity. An atom, with its nucleus at the centre and a number of electrons moving around the nucleus, cannot simply be a miniature of the solar system, because electrons lose their energy by emitting light incessantly until the whole atom collapses in a comparatively short time. Thus, in contrast to the solar system, Rutherford's atom cannot be stable at all, unless it is endowed with some new property alien to ordinary material bodies on the human scale. As I just mentioned, an atom cannot be stable because of the incessant emission of light which was understood as a kind of electromagnetic wave, without any doubt, until the end of the nineteenth century. The revolution in physics started with Planck's theory of the energy quanta of light, as I mentioned earlier; the physicists had been puzzled already by the strange property of light. It was Niels Bohr who succeeded in accounting for the stability of the atom by applying the quantum theory to electrons in the atom. All this is so well known that it is a waste of time to talk about it anymore. I took it up as a good example of both the usefulness and the limitations of analogical thinking. I wanted to point out that analogical thinking becomes all the more fruitful when the dissimilarity as well as the similarity between two things is clearly recognized. In this connection, I would like to say a few words about my own experience. When I tried to understand nuclear forces more than thirty years ago, I came across the idea of taking full advantage of the analogy with the electromagnetic force. However, I was also well aware of the dissimilarity between nuclear forces and the electromagnetic force. Consequently, the theory I reached had been destined to have some resemblance to the quantum theory of electromagnetic fields but to be distinct from it in many respects.

Talking in this way, I have always been conscious of the point of vital importance which remains untouched, that is, "How can a man ever determine the kind of similarity between things that enable analogical thinking to be creative and fruitful?" In order to answer this question, we have to take a further step. When we notice that one thing is similar to another, we admit that they are the same in some sense. For instance, to some of the nineteenth-century physicists like Boltzmann, an atom was the same as a solid elastic body except for the immense difference in size. In this case, the atom was identified with an extremely small solid body. It seems to me that the process of identification not only is the starting point of analogical thinking but also prevails in all kinds of mental activity, so that the analysis of this process may well lead us to the understanding of the secret of creative thinking.

Let us go back to a very simple, but nevertheless a very fundamental, form of identification, that is, what we call pattern recognition. Man acquires the ability in childhood to recognize such simple patterns as squares, circles, and triangles. Such an acquirement is prerequisite for the understanding of Euclidean geometry. It is often overemphasized that the Euclidean geometry we learn in school is a system of statements which are linked with each other through the procedures of formal or deductive logic. But if we remind ourselves how we could understand geometry at all, we are easily convinced that it was not the understanding of formal logic itself but the preacquired power of pattern recognition that enabled us to comprehend what the teacher taught us at the very beginning. For instance, the congruence of two triangles under certain conditions was evident to us, not because of logical necessity, but because we could imagine the movement of one triangle so as to overlap the other triangle completely. The comprehension of formal logic as such comes afterwards.

I would like to point out that man acquires an amazing capacity for pattern recognition without being taught in school, whereas man has to learn formal logic in school. I would like to repeat that we came to comprehend formal logic because we had been unconsciously acquiring beforehand the ability to recognize patterns in childhood, as exemplified in the case of learning Euclidean geometry.

In this connection, a very interesting experiment was carried out by psychologists. A man who was blind at birth had an operation so that he could see for the first time. The psychologists tried to follow his line of vision carefully when he was looking at a triangle in a dark room. At the beginning the line of vision moved randomly, but after repeated practice it moved more closely along the periphery of the triangle. This practice ends up with the very faithful motion of the line of vision along the periphery. This means that the identification between the given pattern and the locus of the line of vision has become complete and definite. Thus the man acquired the capacity for pattern recognition, at least for such simple patterns as a triangle.

We who are endowed with eyesight at birth must have also acquired the capacity for pattern recognition in a similar way in our childhood, although we do not remember it. However, the same psychologists tell us about the result of an experiment which is even more striking and interesting. They tried to follow the line of vision of a normal adult. The motion of his line of vision is very much simplified compared with that of the formerly blind man recently operated upon. The restored line of vision of the latter totters along one side of the triangle and when it comes to the corner, the motion is ended. In this simplified way he can recognize the triangle. How is it possible at all? The information which the man obtains by moving his line of vision is twofold. One is the sharp and clear perception of a small part of an object to which the line of vision is directed. This one may call quantitative or digital information. The other is the more obscure but wide-angled perception of the larger part of an object surrounding the small part on which the vision is focused. This may be called qualitative or analogue information. These two kinds of information work together so as to abridge the process of pattern recognition to a great extent. In this way the process of identification in general and the process of pattern recognition in particular are going on their normal course, one that man does not notice while he is growing up.

The more we think of man's power of pattern recognition the more amazed we become. Consider how one recognizes somebody with whom he is well acquainted. One can instantaneously recognize his acquaintance even in a big crowd. This is certainly a kind of process of identification in the sense that the present perception of the figure of a man is identified with the image of him in the memory. Evidently these two are not exactly the same, but are different from each other in various respects. The man may be wearing a summer suit, whereas he appeared more often in the past wearing thicker cloth. He may look older than he did last time. All this does not disturb the identification. He is essentially the same man as the one in the memory. This we do not doubt. If we go into detail we must ask, "What is essential in identifying two things which are not exactly the same?"

Here the problem of intuition versus abstraction arises; a problem which I discussed to some extent on the occasion of the First Athens Meeting in 1964. I emphasized two points at that time. Firstly, man's ability to abstract was crucial for the creation of an exact science like physics. The appearance of a type of genius such as Pythagoras, who could grasp natural laws in terms of simple and definite relations between numbers, and Democritus, who conceived the invisibly small atoms and the abstract concept of void, was essential for the initiation in ancient Greece (and not in any other land) of natural philosophy which developed into modern science as we know it today. The second point was the importance of man's power of intuition. As a matter of fact, abstraction cannot work by itself, by its very nature. One must abstract something from something else which is more concrete and richer in content. In other words, man has to begin with intuition or imagination, and then he can proceed with the help of his power of abstraction. If we look at the present state of affairs in contemporary science, particularly in fundamental physics, we notice that there are two related tendencies. One is the ever growing tendency to build larger and larger machines such as accelerators in order to carry out creative experiments, as I mentioned earlier. The other is the overwhelming tendency among theoretical physicists working in the field of elementary particles to make use of highly advanced and very abstract mathematics.

We physicists are all very well aware of the inevitability of these trends. Nevertheless I am not very happy about them. The main objective of a natural science like physics is the recognition of nature, which can be deepened and widened by the continuing efforts of scientists. More and more scientists have been cooperating in this great undertaking of man, yet the recognition of nature continues to be a matter belonging to each scientist as an individual. It is true that the physical objects and the physical laws being dealt with in physics today can best be expressed in terms of very abstract mathematical symbols. The question remains, however, as to whether we have to be satisfied with these very abstract things or may we ask for something more? This is not an easy question. There can be answers differing from each other depending on the scientists who try to answer the question.

A physicist belonging to the younger generation is apt to care only for the agreement or disagreement between the experimental results obtained by big machines and the conclusions obtained solely by very abstract mathematical reasoning. This is understandable. Actually, theoretical physics has been pursuing more and more abstract formalism since the advent of quantum theory and the theory of relativity. Thereby intuitive pictures cherished by nineteenth-century physicists have become outmoded one after another. Yet my answer is different from the one just mentioned in the following two respects.

Firstly, as I pointed out earlier, abstraction cannot work alone. There is always an interplay of intuition and abstraction in any fruitful scientific thinking. Not only is something essential to be abstracted from our rich but somewhat obscure intuitive picture, but it is also true that a certain concept, which was constructed as the fruit of man's ability to abstract, changes very often in the course of time into a part of our intuitive picture. From this newly constructed intuition, one can go on toward further abstraction. An example of this from modern physics would be the four-dimensional space-time world of Einstein's theory of relativity, which was very abstract compared with the space and time of Newtonian dynamics, but is today a part of the intuitive picture of physicists which serves as a ground for further abstraction.

Secondly, we must not forget that abstract mathematical formalism is always an end product of scientific thinking in which intuition plays a more important role than is usually noticed. In order to elucidate this last point more clearly, let us go back again to man's ability to recognize patterns, which is a typical form of intuition. Perhaps it would be most useful to compare man and man-made high-speed electronic computers with respect to the capacity of pattern recognition. Certainly one can construct a computer which can recognize simple geometrical patterns. It can identify alphabetic or Arabic figures if they are printed exactly according to prescribed rules, but it cannot identify letters casually written by man. Man's ability to recognize patterns is far more flexible. It is not necessary for man to draw geometrical figures precisely with ruler and compass in order to prove a certain theorem in Euclidean geometry. Man's faculty of pattern recognition evolved to the stage of constructing a general concept of the triangle, for instance, without referring to any one triangle actually drawn precisely. This may also be regarded as a result of the cooperation between intuition and abstraction.

I have already observed that the process of identification underlies what we call the recognition of something. Thus the development in man's intelligence can be looked upon as a kind of evolution of the process of identification. The most naïve form of identification can be seen when a child is identifying his toy automobile for instance. The present perception of the toy automobile is identified with the memory of the toy which was acquired an instant ago. The identification process goes on incessantly in such a way that the child thinks that the toy is moving here and there. The most highly evolved form of the process of identification can be seen in the discovery of the laws of nature by a great scientist. There is a famous legend that Newton came across the notion of universal gravitation when he chanced to see an apple falling. It is said that he asked himself why an apple falls to the ground while the moon does not. I do not know how much truth there is in this story, but it can be regarded as a typical example of analogical thinking. However, Newton's creativity operated at a level higher than mere analogy. What he discovered was the law of nature prevailing not only on the earth but everywhere in the solar system. If there is some truth in the legend, it is the conception by Newton of something essentially the same in the motions of an apple and of the moon. What was to be identified, however, turned out to be rather abstract and general in nature. What was to be discovered was not the identity of the matter which constitutes the moon with matter on the earth, but the identity of the laws which govern the motion of the moon with those which govern the motion of a body on the earth. Thus Newton succeeded in finding out the universal laws of motion together with the universal laws of gravitation.

In short, he did not identify one thing with another, but identified the relation between things in one case with the relation between things in another case. As I mentioned earlier, Newton seems to have been concerned also with the identity of things, which led him to believe in atomism. Perhaps, though, he thought it was premature to deal with atoms systematically.

Now, we physicists know that the laws of motion of invisibly small atoms or electrons turned out to be quite different from the laws of motion of visible things which had been discovered by Newton. We know now the identity of the fundamental constituents of matter and energy. Any two electrons, for instance, are identical to each other, there being no mark whatsoever to discriminate one from the other. If we further ask why they are completely the same, the answer may be that they are created according to the same laws of nature. In this sense, the laws of nature play the role of molds for creating various kinds of particle which are identical to any other particle of the same kind, because they are the products of the same mold. In contrast to the case of Newton, there is no longer a clear distinction between the identification of laws of motion of different things and the identification of the fundamental constituents of matter and energy. The process of identification must evolve to the stage in which the unified description of elementary particles of all kinds is achieved in the sense that both the reason for their existence and the reason for their specific behavior are explained by the same fundamental laws. This is the great expectation of physics. To realize this expectation, the creative thinking of physicists is needed today as much as in the past.

3 The Conception and Experience of Creativity [1963]

The Manifestations of Creativity

Ever since I reached the age of fifty or so, I have been considering the question of how not only I myself but younger research workers also can best display creativity, and have been trying to examine this question of creativity from a rather more objective viewpoint.

I have doubts, of course, whether something such as creativity can be dealt with theoretically at all. The very attempt to consider objectively such a thing as creativity involves from the very outset a contradiction of a kind. Creativity implies discovering something that nobody has hitherto known about or inventing something new. It would be distinctly odd, in fact, if one could say: this is the nature of creativity, so this is what one must do to display it.

Even the research worker striving to bring his creativity into play, does not know, when he feels that it has manifested itself, just how he achieved it. He has the feeling, rather, that something unexpected even to himself has taken place. Invention and discovery always have something of the nature of the unpredictable.

Even so, one still wonders whether there is not some way of making the process easier—of increasing the possibility, of enlarging the probability, that creativity will show itself.

In my own field of physics, creativity comes into greatest play when someone discovers some new natural phenomenon or some new fact, or when someone discovers a new principle, a new law of nature, thanks to which our understanding or perception of natural phenomena, makes a great leap forward. Our comprehension of nature as embodied in natural laws develops into a theoretical system embracing those laws, as a result of which a comparatively wide range of facts can be understood as an integral whole.

Seen in this setting, the new fact that is discovered must not simply be new but must have importance also. The type of discovery that is extremely important is the kind that involves some fact that is most obviously at variance with the laws we know already, or that is extremely difficult to explain within existing systems.

For example, the Michelson-Morley experiment (1887) was designed both to demonstrate the existence of the special frame of reference called the ether frame, and to determine the motion of the earth with respect to that frame.

The expected effect (to be visualized with an interferometer invented by Michelson) was not found, even after a series of experiments of increasing accuracy.

These findings captured the attention of the entire community of physicists, and led finally to the abolition of the long-cherished idea of ether by Einstein who established a new theory of relativity in 1905.

The discovery of new ideas in this sense is extremely important. In this kind of case, the "creativity" is something that comes from without. Here, it is obviously determined by the nature of the natural world itself, working to break down all kinds of existing ways of thinking and fixed ideas.

The Breaking Down of Fixed Ideas

This means that there is also something that needs breaking down within ourselves. The need for reform is not without, but within.

Einstein's discovery of the principle of relativity was unquestionably a major manifestation of creativity, but this does not mean that Einstein was possessed of the theory of relativity from the time he was born. It was contact with negative experimental results contradicting the ether hypothesis, among other things, that made him change his own way of thinking. This means, to put it differently, that prior to creativity there has to be a struggle with one's own self.

A considerable period of preparation is necessary before a particular man can display creativity in a particular field and in a particular form. He must, in short, have acquired all kinds of knowledge and also, probably, have undergone all kinds of training. It is only after many kinds of prior conditions have been satisfied that creativity can show itself. By the time one has done research for a long continuous period and become a full-fledged research worker, one has developed within oneself a relatively stable system of knowledge. This system of knowledge has been integrated by one's own efforts into a particular, definite form. And this business of integrating by oneself is, of course, an extremely valuable experience in itself. It means that one is able to teach others, and to pass on one's own knowledge.

That same state of affairs also means, conversely, that one has become set in one's way of thinking. To exaggerate a little, one has become a mass of fixed ideas. Anyone who carries on his studies for a long enough period of years—myself included— becomes such a mass of fixed ideas.

To know a lot of things has the advantage that, in theory at least, it serves as a basis for discovering new things; but it also has a gradual immobilizing effect. Whatever happens, nothing surprises one; and the chances for a display of creativity are lost.

Drawing Out Latent Ability

In a very large number of cases, creativity is judged by its results.

Einstein, for example, discovered the principle of relativity.

Even those who do not know what it is all about are convinced that Einstein was an exceptional genius and that an exceptional display of creativity took place.

When one reads in his biography that he showed no particular brilliance when he was young, or that he failed in some examination or other, one's admiration for Einstein increases still further. One would be much less impressed if he had always been at the top of his class. It is much more flattering to many people's egos to think that someone who subsequently made a great discovery had flunked his exams at least once. But if he had not made his great discovery, if he had not become a famous scientist, then the verdict, to the end, would have been that he had never cut much of a figure right from his schooldays.

To judge by results is perhaps only natural, but this will not help to elucidate the essential nature of creativity. One must consider, rather, why there was such a manifestation of creativity, what the state of affairs was before its manifestation, and where it had lain hidden all the while.

It is not something that appears out of the blue. Heredity, environment, and the like doubtless all play their part, but however one cares to express it, the most important thing is that there has always existed the possibility of such creativity appearing, that something hidden, something latent, should appear, should become manifest.

Thus the question of creativity, I feel, can ultimately be reduced to the question of where creativity lies hidden, and of the means whereby it can be brought out into the open.

Geniuses Appear in Batches

The seventeenth century saw the emergence of a large number of geniuses. Within the space of one hundred years, an extraordinary number of geniuses—surpassing geniuses, one might say—appeared, beginning with men such as Bacon, Galileo, Kepler, and Descartes and carrying on to Newton and Leibniz.

The beginning of the twentieth century is another case in point, since it produced within a short space of time men such as Planck, Einstein, Rutherford, de Broglie, Born, Heisenberg, Bohr, Schrödinger and Dirac. It is usual, it seems, for geniuses to appear in batches; on the other hand, there are also periods during which the appearance of geniuses is very rare. There must be some reason for this, I feel, other than coincidence.

To take a more everyday example of a similar kind of thing, it is frequently observable in schools that a certain grade or thereabouts will produce a sudden burst of comparatively outstanding young people, which will be followed by a period in which there are none, to be followed again after a while by another similar burst.

There are various reasons for this phenomenon, I imagine, but one that can be grasped very easily is the psychological effect. The presence of classmates who work hard and excel in their studies is an encouragement or a stimulus to the rest to emulate them. This kind of influence probably plays quite a large part.

self-similarity

In the same way it seems likely that on a large scale—over a period of several years, or even of a century—scholars should have a great effect on each other and produce a steady succession of great minds.

The Situation Today

The present age is marked by an extraordinarily large number of scientists. The number of those alive at present probably far exceeds the number of those who have lived and died during the period from ancient times to the present.

If the chances of a given individual manifesting creativity were governed by the same laws of probability as the casting of a dice, the present age should have produced far more great discoveries and great inventions than it in fact has. Why, then, does the reality seem to be otherwise?

I should like here to touch on the situation in Japan, where, if we go back to ancient times, we notice that the higher manifestations of culture were not indigenous but had been brought in from outside.

From two to two and a half millenia ago, for example, Japan's neighbor China produced a large number of great thinkers; there were many systems of thought that still hold interest for us today, and many philosophers whom we may still call great. Japan at that time was scarcely at a stage to produce any organized philosophy. The first appearance of anything that one might call indigenous and systematic thought in Japan occurred with Kōbō Daishi in the ninth century.

This does not mean any intellectual failing on the part of the Japanese, but simply that in ancient times the time lag between the advanced and backward areas of the world was extremely great.

It was because this state of affairs was brought painfully home to the Japanese around the end of the Edo period that they decided to take over Western civilization as speedily as possible.

The influence of such a history on the outlook of the Japanese has in some ways been very persistent. The idea took hold that there was little point in doing creative work at the highest level; that it was better to let foreigners take the lead, and even nowadays reliance on foreign inventions still seems to be too great.

Historically and statistically this is correct. Yet nothing profitable is going to result so long as one deals with the question of creativity in terms of statistics alone.

The basic essential here, I feel, is to break down this set way of thinking. This is surely relevant to the display of creativity. Such things, it is true, just cannot be broken down overnight, but I try to persuade myself that the situation is beginning to change a little, and I go on stressing the need without getting too pessimistic.

No one can say whether a certain person originally had creativity or not. As I have already said, we judge by the results, which are extremely difficult to predict.

It seems unlikely that modern man is inferior to ancient man. Modern man has many advantages over his predecessor. If, despite this, few people display creativity, it means that the mental attitudes of those of us who are alive today are wrong. Mental attitudes, if not everything, are at least one important element—I myself would say the most important of all.

Failure as a Source of Creativity

Even to the best-known scholar, success does not come so many times a lifetime. If one does research for, say, forty years, one is likely to achieve real success only two or three times in all. Some men go a whole lifetime without achieving a single major success. What were they doing all the time, then? Not nothing, of course.

In long years of research, the right opportunity comes along very seldom. Even when it comes, one often fails to grasp it. The result, in short, is that one seldom succeeds, and so the manifestations of creativity are extraordinarily valuable. If success was easily gained, there would be little need to talk of creativity at all. Even the man who, seen from the outside, seems to have had a reasonable amount of success has almost certainly experienced repeated failure. The failures, of course, are in no sense wasted. One's failures prove a later source of success. Nor need it only be one's own failures; sometimes the failures of others prove most significant. What others have tried and given up one alters a little, and one succeeds. As the old saying goes, "Failure is the mother of success."

I myself often work hard from morning till dusk only to throw what I have done into the wastepaper basket in despair. Quantitatively, the amount I keep because something might come of it is incomparably smaller than what I throw away, yet it is this, I believe, that serves as the basis for creation.

Tenacity and Thinking within a Framework

I have suggested above that the reason why so little creativity is displayed despite the large number of scholars and the advantages they enjoy over their predecessors is that there is something wrong in the mental attitudes of the Japanese. Another reason—though this is susceptible to change—is the fact that the environment in which we find ourselves is extremely unfavorable to concentrating one's efforts for long on the same thing.

Looking back over my own life, for example, I find that the number of miscellaneous tasks I have had to perform has gradually increased over the years. Wherever possible I try nowadays not to take on tasks that fall under this heading, but still I have too many.

On top of this, there is too much information. The bewildering influx of new stimuli leaves one no room to think about things in a relaxed way. Sometimes one tends—is obliged—to flit from one thing to another.

What I find particularly awkward is that this information comes to one already arranged and standardized.

When the amount of information coming in becomes excessive, it becomes quite impossible to present it just as it stands. It would be an enormous labor for the invidual to set it in order by himself. In practice, and for better or for worse—for worse, in many cases —it has already been set in order for us. The news in the papers and on the television, for example, is a perfect example of this process at work. For convenience's sake, we accept such things in the form in which they are served to us. To do so gradually becomes a habit; it seems to make life easier, yet it also steadily increases one's dependence on others.

At the same time the very fact of arranging, whoever has been responsible, implies the existence of some method or framework. Even when a person does the arrangement himself, so long as he thinks of things within a fixed framework there is no creativity. All major creation begins with getting outside the framework, or changing the framework itself.

The important thing if one is to achieve creativity, I feel, is to keep plugging away at one thing despite all the miscellaneous tasks and the superfluity of information with which everyday life seeks to claim one's attention. What is needed, in other words, is tenacity of purpose.

Physicists who show creativity usually work with extraordinary —almost unnecessary, one might think—tenacity at a particular subject.

This might seem to contradict what I have just said concerning the need to break down fixed concepts, and it is an extremely tricky point, yet the fact is that most of them have some ideal, image, or vision, which they cling to with unusual pertinacity.

I, for one, have my own particular subject; in the years since I first took up physics, there have been many changes and much development, yet essentially I have been concerned with the same thing throughout.

The question is, to what one should attach oneself in this way? Subjectively speaking, anything should do equally well, from pinball games to scholarship. Yet objectively viewed, it is more meaningful to devote oneself to something as difficult of attainment as possible. One may suffer one's whole life as a result, but it is well worthwhile.

One further question is that the act of devoting one's tenacity to one thing also means, in itself, that one is aware of some contradiction within oneself. Without some contradiction within, there can be no study; that, indeed, is the essential nature of study.

To put it differently, one has some place that is dark, or obscure, or vague, or puzzling within oneself, and one tries to find some light in it. Then, when one has found a ray of light, one tries to enlarge it little by little so that the darkness is gradually dispelled. This, I feel, is the typical process whereby creativity shows itself.

Custom, Imitation, and Creativity

Around 1925, the year preceding that in which he did such important work on quantum mechanics, the great physicist Schrödinger wrote an article concerning his view of the world. In this article, he writes of the consciousness as a kind of outpost of the unconscious, a light that flares up suddenly in the darkness, in the night of the unconscious. This relates directly to the question of creativity, for consciousness is generally connected with some new or fresh experience.

If one goes on repeating the same thing, it gradually becomes unconscious. One comes to do it by reflex, or perhaps one should say by habit. It is the escape from habit of this kind that constitutes creativity.

All kinds of habits are firmly established within us, forming a kind of fixed system, and when the consciousness comes into contact with something that does not fall within that system of habit, it often becomes extremely acute.

Another thing is that men copy each other. It is unconscious copying of those about us that helps us grow into adults. We are obliged to copy in all kinds of ways, otherwise social life becomes difficult for us. This type of imitation is the precise opposite of creativity.

Imitation is the creation of something that is already in existence. When I was a child, I imitated the way my brother, three years older than myself, ate at table. My brother sat directly opposite me. He held his chopsticks in his right hand, so I, thinking I was imitating him exactly, held mine in my left. Once my mother noticed, I changed over to my right hand, but, possibly for this reason, I still wield my chopsticks clumsily even today. I often, in fact, cause amusement by the clumsy way I clutch them in my fist.

I don't mind being laughed at, but what is awkward is when I have a foreign visitor and I take him to some Japanese-style restaurant. Since I am often asked to show him how to use chopsticks correctly, I have to take my wife along; she uses them very skillfully, and I tell him to follow her. Before long, I see that he is holding his chopsticks in the proper way whereas I myself am still clumsy. I feel ashamed, but even if I try holding them correctly it doesn't last for more than a minute.

Imitation seems to be of no use while one goes on and on repeating it. One never becomes proficient. But just occasionally creativity arises out of this very process of repetition.

This kind of imitation becomes, in the broad sense, memory; man has the power of memory, whereby he stores up his experiences. Without this store of memories, creation is impossible. Yet memory means, as we have seen, to store up experiences and re-create them—repeat them—as required. Recalling is, in itself, a kind of repetition, and there seems to be no element of creation in such a process. Just why it should sometimes give rise to creation is still not really clear.

The Validity of Logic

At some stage or other, human beings acquire the power of rational thought. They learn, for example, to distinguish between truth and falsehood, between right and wrong. They learn that one cannot both affirm and deny the same proposition at the same time, and that one must reject one or the other.

As the power of thought in this sense develops, one becomes able to take in formal logic. We are unconsciously imbued with the idea that one must be faithful to formal logic. Yet in practice we are not faithful. Our reasonings are highly suspect; it often happens that by the time we have developed three or four times the thesis that "such-and-such is so, therefore such-and-such is so," we reach an extremely doubtful conclusion. For in most cases, if one concludes that such-and-such is so with a veracity of, let us say, 70 percent, people will allow themselves to be convinced where that particular fact is concerned. If one then proceeds to deduce from this that such-and-such is so, and if this reasoning, in itself, similarly is 70 percent correct, then the truth of the conclusion will decline to less than 50 percent.

Inductive logic, it is often claimed, is unlike this in being extremely reliable, and it is on inductive logic that scientists rely. Yet this, too, is rather shaky. In many ways inductive logic as it is generally known is very prone to cause misunderstanding.

Let us suppose, for example, that one carries out experiments under all kinds of different conditions, collecting data which lead to the discovery of a law. If one carries out extremely delicate measurements, varying the conditions very frequently and subtly so that one obtains a great deal of data from the experiment—if, in short, the data come to speak for themselves—then nothing very special has been achieved in terms of creativity. In practice, the development of science has been achieved less in this way than by a succession of bold sallies involving the discovery of a simple law from comparatively scanty data and from measurements not in themselves so accurate. To put it round the other way, what is referred to as inductive logic here involves something more than faithfully reproducing experience. There is a leap from the particular to the general. There cannot be discovery until that leap occurs.

At this point the question arises of whether what is referred to as a law in such cases is not really in fact a hypothesis. Both in natural science and in psychology, the fact that something is practically demonstrable is extremely important. However, when one attempts to understand all kinds of empirically observable facts as a whole, it is necessary to resort to laws as intermediaries. The "laws" in question here always have the nature, rather, of hypotheses.

When a number of more or less empirical laws are established, physicists tend to find more general laws comprising them, gradually evolving a basic theory. Thus the hypothetical element becomes increasingly pronounced. It is precisely because of this hypothetical quality that demonstration has the decisive importance that it does.

In this connection I stated above that, although we consider formal logic as such to be extremely dependable and are convinced that it is an infallible method of reasoning, in actual use it is not always valid. If, however, one introduces numbers, one can say things with a very high degree of certainty; for the process of demonstrating with numbers, being deductive and formal logic, permits of a perfectly clearcut yes or no, with nothing intermediate possible. However much one repeats the processes of reasoning, thus, there are no dubious conclusions. Numbers are valid, and it is for this reason everybody puts their faith in them.

One way of dealing with numbers, however, is what is known as mathematical induction. To go into details here would create too much of a digression, so the interested reader is referred to Poincaré's *La Science et L'hypothèse*. Very simply speaking, however, it goes as follows: suppose that we have proved that if a statement is true of an arbitrary integer n, then we know that it is true of $n + 1$; and also we have proved that if it is true when n is 1, then it is true of all n's. It is a kind of building up of deductive logic, but first of all the formula in the case of n has to be discovered. Here a creative ability has to be brought into play.

The thing in which mathematics is obviously different from any other branch of study is that in empirical science there must be correspondence with actuality. In modern mathematics, this is not necessary, but in empirical science constant reference to actuality is obligatory.

The question arises here of what is actuality, for scholarship would get nowhere if there were nothing but reality as such; we have to understand it with the aid of some intermediary such as words. In practice, we have no way of understanding without an intermediary of some kind.

Facts alone are just facts; what is necessary in order to get beyond that point is to put various different facts together to produce something new. In order to express the overall structure of a number of different things—though I dislike using the word "structure," since it is used nowadays in so many different senses —one must introduce concepts and laws.

Creativity lies in this business of putting together to produce something new. This is an extremely important point.

On Analogy

A method in use since ancient times as an aid to understanding something difficult is to draw a parallel with something easier. Generally speaking we never, in practice, understand a thing simply in itself; instead, we draw diagrams, or express it in words, or in a numerical formula. In every case, in short, we are likening it to something else. This, taken in its narrower sense, is the analogy. With an analogy, we understand one side very well, but not the other. The thing we do not understand well seems, some-how, to resemble the one we do. We start to consider just how they are alike—and, in a flash, the thing we did not understand is suddenly elucidated. This kind of thing is happening all the time.

In physics, it takes the form of thinking in terms of models. For example, we take the solar system as a model in considering the structure of the atom. Sometimes, in short, we build up an image by means of which we understand the actual object. We have the ability to form such images, to summon up an organized picture. This ability is a fundamental and vital factor in the manifestation of creativity.

The Lack of Metaphysics

Professor Kikuya Ichikawa of Dōshisha University has distinguished two different types of information which he calls "numerical" and "conceptual" respectively. A typical case of the former is that processed by an electronic computer, which consumes numbers and spews out numbers in return. The information it supplies is "numerical" from first to last; at present, it is incapable of creative activity. It can handle information of a considerable complexity, and carry out difficult calculations for us, but it has no creativity of its own.

One of the reasons for this is that the computer does not possess the ability to conceive, or perceive something as an integral whole. It lacks the human faculty of figurative recognition.

As I pointed out earlier, the present age, seen in terms of the creativity question, is extremely badly off; the display of creativity occurs infrequently when compared to the number of scientists. One reason here, I feel, is that everybody has become too "digital."

Unlike mathematics, physics needs to maintain its connections with the actual world in which we live, yet in practice it has become divorced from actual things and preoccupied solely with abstracted theories, experimental data, and the like. The only checking of the facts that is done in practice is to collate experimental data given in numbers and, on the theoretical side, the results of calculating from highly abstract formulae; if the figures on both sides match, people congratulate themselves and that is the end of the matter.

I ask myself just what scientists think they are up to. How can the majority of them be content with such an empty repetition?

To put it in different terms, what it comes down to is a lack, a complete lack, of metaphysics. In the nineteenth century, they used to say that science had nothing to do with metaphysics, which must be rigorously excluded; the business of science was to describe and to demonstrate, and it must concern itself with nothing but facts. A similar kind of argument, in a different form, is to be heard today too. But I cannot believe that this is what human learning is all about: it should spring rather from a more basic desire, from what one might call a desire to discover the truth. I have got tired of reiterating this theme, for my opinion seems to be so much at variance with the general trend of our times.

It's something like the man who looks at abstract art and doesn't find it appealing; it is dishonest for him to force himself to nod his head admiringly. Indeed, I very rarely find a work that impresses me; I am sure there *are* some good works, but I've rarely come across them. It's a great pity; I think there ought to be something more to art than that. However, this is not the time to air my views on art. . . .

Even so, though, if one reads Schrödinger's *My View of the World* which I mentioned a while ago, one finds the same complaint. It is all very well—he, too, wrote—for physics to be descriptive and demonstrative, but if that meant ridding it of all metaphysics he could not but feel a sense of emptiness. He had this feeling many years ago when physics was much less abstract than it is now.

Identification

I spoke a while ago about analogy. There can be no theory of creativity, I feel, that does not take analogy—which is itself one of the functions of the human intelligence—as its starting point. It is here, and nowhere else, that we must look for the essence of creativity.

In actual practice, all the thinkers of the distant past relied on analogy in developing any ideas worthy of the name.

Nowadays, I am inclined to think that the fundamental thing in any consideration of creativity is the function of the intelligence known as "identification." I do not mean to imply that identification is in itself the same as creativity; but I do believe that the power of recognition of two things as essentially identical has a decisive importance within the human intellect.

We see a flower, and we know it is a flower; it is not simply that the flower is visible to us, for we can also recognize it for what it is.

As for the nature of recognition: when I say "this is my watch," I am already using the word "watch" as an intermediary, I already know what is the thing known as "watch." I see this watch of mine every day. The memory of it and the actual perception of it by my eyes are identified.

When one reaches a certain age, one begins to wake up to oneself, to become more aware of oneself; one recognizes oneself as oneself; one identifies oneself as oneself. Paradoxically, it is only after one has acquired this awareness of the self that one acquires a philosophy. There is philosophy even before this, of course, but it can be distinguished from what follows as natural philosophy. One is oneself, one is the same as oneself; that appears trivial, and yet, despite that, the identification of oneself as oneself is the beginning of all sophisticated philosophy.

A Manifestation of the Life Force

One of the characteristics that distinguishes the living thing is reproduction. It has the ability to reproduce, to go on recreating, something that is the same, or very similar to, itself, something that even when slightly different is essentially the same.

In recent years, it has become clear that an extremely complex molecule known as DNA is the basis of all heredity. This DNA molecule has the ability to produce other molecules the same as itself. Thus if a certain DNA molecule comes into existence it will, given the right environment, create something the same as itself.

The probability of something so extremely particular as a human being coming into existence is basically very small, but once it has come into existence it has the ability to reproduce itself—in other words, it has the mechanism for reproducing itself. Since what occurred by chance can create a large number of things the same as itself, chance ceases to be chance any longer.

Thus one may say that within the basic mechanism of the natural world in which we live, life is also, in itself, a manifestation of a kind of process of identification.

More fundamentally still, there is, at the very bottom of the natural world, something like a kind of invisible mold. This invisible mold is, expressed differently, a law. It produces identical things. It can, for example, produce infinite numbers of electrons each differing in no respect from the rest. That is the essential nature of the mold, and in the reproduction of precisely identical objects we have a process of imitation that is precisely the reverse of creativity in the sense in which we are discussing it here.

If that were all there was to it, there would be no question of creativity, but in the world of living creatures there is what is known as sexual reproduction, in which a new combination is made from the DNAs of male and female. There is also the sudden mutation, where for some reason or other—whether what appears to be chance is really so or not—something different appears suddenly. Where this sudden mutation has the capacity for reproduction, it too goes on increasing indefinitely. At the level of living creatures in general, one may see this process as fortuitous, but in the case of human beings the same kind of thing is done consciously. Human beings consciously produce things that are different. The things that are produced consciously, however, develop, in a different way from living things, an extremely great potential for reproduction.

Such is the case with learning. Tell someone a piece of scientific information, and he perceives the truth of it; in this way, knowledge spreads ever wider and wider. Publish the information in a book, and people read it and are convinced in their turn. An enormous power of reproduction is at work here.

Thus conscious effort by human beings at the human level can be seen as an extremely advanced manifestation of the life force, and if this is so then the tenacity that I cited earlier as an extremely important basic factor in the manifestation of creativity can also, I think, be seen as a manifestation of the life force in a very broad sense.

Not only things and energy, but all things that actually exist—time, space, our own minds even—are at the same time constantly changing. Changing, becoming—by, for example, the birth of something new—is of their essence.

Fundamentally, this fact is linked up with the way human beings are forever creating new things; thus the display of creativity by a human being is, in the same way, a manifestation of his vitality, of the life force.

Tenacity of purpose is also, at bottom, a question of vitality.

For such reasons, I am increasingly inclined to see the intellectual ability to identify, which all human beings possess, as a clue to understanding the nature of the world, human beings included; and I am hoping to construct from it some organized theory of creativity.

Even if I should succeed, however, I do not persuade myself that it would provide a universal key to the display of creativity. Things such as creativity just cannot be dealt with in terms of theory alone; without the element of personal experience, in short, it will be flat and lifeless.

However long I go on talking, I seem to reach no final conclusion. If we meet again, I may be able then to present you with a better organized and more advanced theory of creativity, but I hope you have got at least something out of what I have said today.

IV ON THEORETICAL PHYSICS

1 Facts and Laws [1947]

The Standpoint of Common Sense

Science is constantly on the move, a fact that also signifies that science is never quite complete. This special quality of science derives, of course, from the fact that it starts with experience and, since the possibility of fresh experiences is forever opening up before us, returns to experience in the end. These things that we label vaguely as "experiences" are extremely complex and extremely varied, yet we know that they are not in fact a chance jumble but are interrelated in all kinds of ways. The ability to follow up, or to discover, their interrelationships depends, it is generally agreed, on the gift of reason. In this respect, Galileo's statement that science depends on experience and reason is one of the eternal truths.

This process of tracing the relationship between our experiences can be divided into two parts: the ascertaining, in one sense or another, of the empirical facts, and the establishing of some law that shows the relationship between a large number of facts. However, when it comes to the question of what constitutes a fact, and what constitutes a law, then not only will the meaning of these terms differ according to the precise point reached in the development of science, but it also becomes plain that there is in fact a close connection between the two questions. In particular, the rapid development of physics since the beginning of the twentieth century has had a great influence on our thinking concerning facts and laws and the relationship between the two. In a sense, one might almost say that to define these three things accurately would be, ultimately, to elucidate the essence of science itself.

Before delving deeply into this question, however, it will be as well to start by setting forth the "common sense" approach, the approach that seems self-evident to everybody. Doubtless our various experiences are related in all kinds of ways, but the first thing that will strike anyone is that they follow a certain order in the way they appear, pursue their course, and cease. Almost unconsciously, we recognize the existence of something called "time" that forms an undercurrent to such changes, though it might be truer to say that we see the changes, in themselves, as time's method of self-expression. Essentially, then, time might seem to be an extremely subjective thing. In fact, though, what we understand when we use the word "time" is something different. Unconsciously each of us is convinced that there is one single stream of time that is not private to ourselves but shared by all human beings alike. And we agree that this "absolute time" does not necessarily reveal itself directly, in the order in which we experience things, but that in many cases, for a variety of reasons, an event that chronologically speaking should be situated at a very early stage is not experienced by ourselves until much later. Let us suppose, for example, that perusal of a newspaper today tells me that a Mr. A died yesterday. Even assuming that, so far as my own experience is concerned, the demise occurred after breakfast, I am obliged—provided, that is, that I trust the newspaper—to conclude that it occurred before. In this way we are all, constantly, rearranging our experiences into a particular natural order. By doing so, we assume almost unconsciously that the order is something common to all men and unrelated to our own subjective nature. Indeed, this conviction is adequately borne out by the fact—in itself, admittedly, an *experienced* fact—that if we compare individual judgements concerning the natural order, they always correspond with each other. In such cases, it is usual for us to turn for confirmation to "absolute time" as represented by the position of the stars, or of the hands of some clock that can equal them in accuracy, in order to determine the order accurately. If, for example, I ate dinner yesterday at around 6 P.M. and Mr. A died at 7.21, then I may assume that so long as the clocks were not seriously wrong, my eating of dinner occurred first.

The objective significance thus attributed to chronological order is a prerequisite of a causal relationship—one independent, to a certain degree at least, of our subjective emotions—between various experiences. Assuming that I also knew Mr. A to have been ill in bed for the past month, then it is obviously a necessary condition for any inference concerning the connection between the two facts that their relative position in time should be a fact independent of my own subjective feelings. It is also clear, though, that chronological order alone is insufficient to establish a causal relationship. Even though I may have been eating dinner a short while before he died, it hardly seems likely that there was any cause-and-effect relationship here. To single out the illness as the cause of death from among all the things that happened to him preceding it implies a fairly strong reason for so doing. If the illness is one that is known to be difficult to cure, then the assumption that it was the cause of death is very convincing indeed. If it was, say, a cold, or intestinal catarrh, then we suspect that though it may have been a contributing cause there were other, more important ones. In most cases, the thing that we refer to vaguely as the cause of an event is in fact only one of the causes. One cannot, thus, deduce the inevitability of the result from it alone; all one can do at the most is to infer the probability of a particular outcome proceeding from some cause or other.

Does it mean, then, that if only one uncovered all the causes one could fix on a single effect? If, at the time Mr. A fell ill, one had had adequate knowledge of his physical condition and the circumstances in which he was placed, could one have foretold that he would die in one month's time? It is difficult to give a definite answer to such a question on the basis of common sense alone; even a specialist, much less a layman, would have difficulty in predicting the time of a man's death one month in advance. However grave the illness, it is still risky to assert that the patient has absolutely no chance of recovery. This difficulty in establishing a primary connection between cause and effect is, in fact, the rule rather than the exception. Is this because of the inadequacy of our knowledge and experience, or is it impossible in principle to foretell accurately future events however complete our knowledge and experience? To find the answer to such questions of principle one can no longer rely on common sense, but must have resort to science or to philosophy grounded in science.

Before going any farther in this direction, however, there is one more question that can be more or less understood via the common sense approach. Almost unconsciously, we usually divide our many and varied experiences into two groups, one relating to the outside world and one to our own minds. When I am looking at a flower, for example, I am convinced that there is a flower in the world outside myself. At the same time, I am equally convinced that the feeling that the flower is beautiful proceeds from my own mind. Whatever the experience, so long as it is mine, there is always some part of it connected with what I think of vaguely as "my mind"; at the same time, I am obliged to acknowledge that it is also connected in some way or other with the external world, whether in the present or the past. Even when one is happy or sad for no immediately apparent reason, it seems clear that something more than one's own mind is involved, and that there is some close connection with the circumstances in which one finds oneself. Even in the case of dreams, where one realises on waking up that what had seemed to be related to things in the outside world was in fact simply a working of the mind, one cannot deny the connection between the content of the dream and various stimuli from the outside world received before going to bed or while asleep—though in this case the relationship may be too vague and complex to be clearly apprehended.

Everybody, then, acknowledges vaguely that there are these two parts to our experience, but all kinds of difficulties arise when we attempt to distinguish clearly between the two. Suppose, for example, we see a flower and tell ourselves that it is red. We are convinced that—provided they are not blind or color-blind—the redness of the flower is apparent to other people as well. In other words, we believe it to be a property of the flower as an object in the external world. In the same way, the scent of the flower, we are sure, is apparent to everybody else's olefactory sense as well as our own. This too, we believe, is a property of the flower in the same way as its color. Similarly, we agree that the flower that possesses these qualities possesses a fixed shape, that it is connected in a particular way to the stem and leaves, that the whole has been placed in a vase, that the vase stands on a table, and so on. This means that we recognize flower, stalk, leaves, vase, and table as all existing in a single world. More precisely, we recognize that they all occupy their own positions in the same space and at the same time. The part of our experience that we refer to as related to the external world means, in short, the part that relates to various objects arranged within the same space-time. That they should exist together at the same time, occupying fixed positions relative to each other, and with fixed shapes, gives a marked degree of order and stability to the welter of experiences with which we are faced. It implies a kind of "simultaneous relationship." We are tacitly agreeing here that the space in which the objects are situated is independent from each of us in the same way as the objects themselves, that it constitutes a single, "absolute" space common to all human beings.

Thus our experiences are, on the one hand, constantly being rearranged in reference to absolute time, while on the other hand they are arranged still more precisely by reference to various objects that we believe to coexist at the same point in time. This process of arranging is rounded off finally by our assumption that these objects are disposed in absolute space in accordance with a natural order. This means, in turn, that any change in our experience is referred to some change in the disposition of those objects occurring with the passage of time. It is clear from this that in order to put our complex experiences in some kind of order it is most convenient to relate them to the movement of objects. I hardly need to point out here that from early childhood it is to moving objects that our attention is drawn most strongly, it is only later that the concepts of time and space are clearly formulated.

So far, then, there is no problem. Unfortunately, the rearrangement of experience does not end here, but forges ahead under its own steam, as it were. We are obliged to acknowledge that, in the same way as the flower and the table, the bodies of other people looking at them also occupy their own places within the same space. And we are also obliged to acknowledge that our own arms and legs, in the same way as those of others, have a fixed form and color. Though they are parts of our own bodies, insofar as they, too, are one type of object situated in space, they differ in no respect from the bodies of others, or from the flower. One obvious difference between us and the external world is that we are alive. But if that is all, then cats and dogs are surely equally alive. Even the flower that we have treated as an object pure and simple is, in fact, part of the body of a living plant. To all appearances, there is little difference between the flower as it was growing in the ground and the flower picked and placed in a vase. The distinction between the living and the nonliving cannot be determined so easily by common sense.

A second difficulty is that we human beings are not merely living, but have minds too. That the mind is not an object is obvious, yet it is inseparably related to our bodies. The men of old had vague ideas of the mind as situated in the chest or in the belly. Modern man believes that it is situated in the head. But no one can say with accuracy just where and in what form it exists in the head. Certain philosophers would probably reply that they did not believe the mind to be in the head; the mind itself—they would say—has no form and exists nowhere in space; to locate the mind in the head means merely that the workings of the mind are carried out in the brain. It seems to me that such an answer, however, is itself no more than a kind of common sense. The word mind is used in all kinds of ways, but this does not mean that there are all kinds of minds; rather it signifies that there is one extremely complex and subtle thing that is capable of all kinds of interpretations. Whatever the case, what cannot be denied is that, while to some extent it transcends the space that is the constant background to physical objects, it is also closely connected with our bodies which are situated in that space.

A third stumbling block is the existence of a concept still more abstract than that of the mind. We are convinced that there is something known as truth. This is seen at its purest in the theorems of formal logic and mathematics. For us, the fact that two and two make four is a truth that admits of no doubt. Truth undeniably exists; yet it could not be said to exist anywhere particular in space. If one takes two cakes out of each of two boxes, the total is four. Yet though one might say that this is an illustration of a mathematical truth, the fact that two and two make four holds good independently of the time and place where the action of taking the cakes out of the box occurs. This is referred to as an eternal and universal truth. Now, we did not know this truth from the outset. We discovered it during the process of consciously or unconsciously ordering our experiences. The discovery of the truth is in itself, in a sense, an experience. The truths we discover are generally speaking in no way isolated things. A "mathematical truth," for instance, always invariably occupies its place within a system of mathematical propositions free from contradictions. The theorem that the sum of the internal angles of a triangle is equivalent to two right angles has a universal significance over and above the description of empirical fact only because it finds its own place within the system of Euclidean geometry. In other words, it can be deduced logically from a group of other axioms. The reasoning here follows a fixed order. Although this order is distinct from the absolute time that we introduce in order to rearrange experiences related to objects in the external world, one can distinguish here a kind of abstract and noncontinuous chronological order. While a particular theorem evolves from within this ordered process of logic, it often happens on the other hand that it is brought into conflict with other propositions that are contradictory to it. To take the theorem just mentioned as an example, a conflicting proposition would be that the sum of the internal angles of a triangle is not equivalent to two right angles. So long as one accepts the axioms of Euclidean geometry one cannot admit such a proposition into the system of geometry. Yet this does not mean that it is an absolute fallacy. If, for example, one adopts the axioms of non-Euclidean geometry, one is obliged conversely to accept this proposition as representing the truth. This implies the possibility of a number of different systems of logic existing independently—though frequently to the mutual exclusion of each other—at the same time. Although they do not of course coexist in space in the normal sense, one can nevertheless postulate a kind of extremely abstract and noncontinuous chronological system. The next question that arises is whether it is possible to find any relationship at a higher level between the ordering or arrangement of these more or less abstract concepts or propositions and the temporal and spatial relationships existing between our concrete experiences. This is one of the most basic questions facing science.

The Approach of the Exact Sciences

It is said that the special characteristics of science are its empiricism and its rationality. Viewed negatively, rationality implies the absence of any logical contradictions in the reasoning, but positively viewed this is not enough; what is required in addition is the establishment of some system with a high degree of uniformity and of conformity to laws. With empiricism too, a negative definition would mean merely that correlation with experimentally observed facts is always used in evaluating the results of reasoning, but one should not overlook the positive side, either: the discovery of a new conformity to recognizable laws through the furnishing of ever fresh empirical facts. So far, there is no particular objection. The question lies in the precise meaning of the terms "empirical fact" and "law" that are such important concepts for science. I have already described above how the part of our experience that is related to the outside world is reordered spatially and temporally, being referred to various objects disposed in the absolute space shared by us all. I have also described how it seems to us that changes in our experience can be related to changes occurring with the passage of time in the relative position, form, and other qualities of those objects. In practice, however, whereas the position and shape of objects, together with the movement that arises from changes in these, can be described accurately within the framework of space and time, other qualities such as their color and smell are inevitably ambiguous in the extreme. It was not

without reason that Locke distinguished between primary characteristics of objects, such as movement, size, and shape, and secondary qualities such as color, sound, smell, and taste, If we wish to
give some accurate and objective significance to those parts of our
experience that appear to be related to the external world, we are
obliged by some means or other to relate them to the primary
characteristics of objects. The establishment of the Newtonian
system of mechanics concerning the movement of objects encouraged the idea that this was in fact a profitable way of tackling
the question. And in practice, as a result of subsequent developments in physics and chemistry, sound came to be ascribed to the
vibration of air, light to the vibration of ether, and smell and taste
to the contact of special molecules or ions. In such ways, it came
to seem increasingly likely that all the physical and chemical
phenomena that account for so much of our experience would be
related to the system of mechanics in the broad sense. And from
some time after the beginning of the nineteenth century it did in
fact begin to seem that scientists had no alternative but to accept
the mechanical view of the world that held that it should be
possible, by relating all our experiences to the movement of objects within the framework of absolute time and absolute space,
to demonstrate a strict causality at work in the background of all
phenomena.

At the same time, however, a number of awkward questions
had already arisen that foretold future hitches in this simple and
arbitrary trend. If, for example, light was a result of vibrations in
the "ether," then ether must be very different from ordinary ma-
terials such as metal, water, or air. The appearance of the theory
of relativity and the denial of ether's existence in the twentieth
century is too well known to require detailed description; what
happened, in essence, was that it became impossible to treat
"ether" as a material existence in the framework of absolute time
and absolute space. The development of the quantum theory and

quantum mechanics made it still more obviously impossible to consider atoms, electrons, and the like as miniparticles moving along particular courses in the same way as ordinary objects. In order to explain rationally all kinds of phenomena associated with light, atoms, electrons, and so on, it was necessary to look to the theory of relativity and quantum mechanics—and here it was no longer possible to deal in terms of things such as absolute space, absolute time, or objects within them possessing particular shapes and moving in particular ways. We are obliged nowadays to deal in terms of a four-dimensional space-time, and of abstract entities such as electrons and photons that have the dual properties of waves and particles, which could not be combined together in the framework of nineteenth-century physics. In such ways the development of physics in the twentieth century had led us away from the material world, set in the framework of absolute time and space, that was until recently the only reasonable image of the external world, and had shown us another and totally unexpected world.

This new world is far from being a mere fancy of the physicists. Unless foot is set within it, it will no longer be possible to clarify the laws linking together empirical facts over a broad field. The question again arises here, however, of what exactly constitutes an empirical fact. To look at a flower unquestionably constitutes an experience. One empirical fact that emerges from this experience is that there is a flower there. The flower is patent, we presume, to everybody, and if necessary its location and shape can be determined as precisely as one may wish. Beyond doubt, it is a "fact" fit to be dealt with by an "exact science" such as physics. The pleasant smell emitted by the flower, however, cannot be regarded as a fact subject to the same degree of precision as the shape that is seen by the eye. Not merely is the smell apparent only to those close to the flower, but the quality and strength of the feeling are very vague, and are unsuitable as objects of exact scientific study. The color of the flower is considerably less ambiguous, but merely to cite differences such as "redness" or "blueness" will hardly serve as an accurate quantitative description. This is why the physicist has reduced the qualitative differences between colors to numerical values, by postulating waves of light each with a fixed wavelength. In this case, we intuitively think of waves of light being transmitted through normal space in the same way as waves of sound. Strictly speaking, of course, one cannot express these "waves of light" accurately and objectively without plunging into such abstract concepts as the distribution of electromagnetic fields in four-dimensional space-time; but, for practical purposes, there is no need to go to such lengths in most cases. Supposing, for example, one breaks light down into a many-colored spectrum by means of a prism, then the measurement of the wavelengths of light of each color boils down to determining the position of the spectral lines on the photographic plate. And the position in this case means the position in the familiar, everyday, absolute space.

If one uses, say, the gradations on the dial of a galvanometer to measure accurately the brightness of an electric lamp, one is in a similar fashion making up for the inaccuracy of the human eye's feeling for strengths of light by relating them to a visible object—in this case a solid object such as the needle of a meter. This would seem to mean, then, that the world of facts dealt with by the exact sciences is nothing other than the whole range of changes that occur in the position or shape of visible objects within the framework of an absolute space and time. And insofar as the changes in the position and shape of those objects are laid down by our old friends the Newtonian laws of motion, we have arrived safely back in the world of classical mechanics. This is surely most odd: why in that case should physicists have taken such pains to think up difficult theoretical systems such as the theory of relativity or quantum mechanics?

The answer is, in fact, already provided by the explanation just given. Briefly, it comes down to this: our experience is incomplete; and as we supplement it with our reason, we are obliged gradually to expand our thinking, until finally we reach the theoretical systems of modern physics. This answer is so simplified, however, as to be almost unintelligible, so I will be more specific. As a result of setting our complex and varied experiences in order and extracting from them facts that can be dealt with by the exact sciences, we have reached a kind of world of facts, in the narrowest sense. The world we have reached is the world of classical mechanics, and by relating our experiences to events in this world we give our science a convincing empiricism and objectivity. Nevertheless, where the rationality of that science is concerned, things are very far from satisfactory. It is true that the needle of a galvanometer may follow the Newtonian laws of movement. But in order really to determine the movement of the needle, one needs to know the right-hand side of the Newtonian equation—to know, in other words, just what is the force at work in the needle. If one tries to trace the source of this force, one is obliged to take leave of mechanical phenomena in the narrow sense and expand one's field of consideration to include electromagnetic phenomena. When one wishes to measure extremely accurately the strength of a source of light by means of an electric current the question is still more difficult. Let us consider, for example, the detection of an extremely weak light by enlarging a minute electric current flowing through a counter. The ideal thing here is something like a "photon counter," in which there flows an electric current that permits the measuring of a single photon striking the photoelectric counter.

With such an extraordinarily sensitive piece of equipment, if one wishes to find the reason why the needle of the galvanometer moves, one must trace it back to the microscopic process whereby one atom in a light source near the photon counter emits a photon.

In practice, however, the process whereby an atom emits a photon follows the laws of quantum mechanics. Once one reaches this point, the classical mechanics is completely invalid. Here the question is not the continuous movement of electrons within the atom but the probability of their jumping from one state to another state. To relate this back again to the world of facts, it means that the movement of the galvanometer needle is controlled by the laws of probability governing a microscopic process. The microscopic world has its own laws. For example, the shape of the wave function that seems to govern probability changes in a particular manner with time. Yet no one could deny that in the world of facts it is associated with a certain type of uncertainty. So long as the moment when a photon is emitted from an atom can be foreseen only in terms of probabilities we can hardly forecast beyond a certain point when the needle of the galvanometer will move. To look at it from a different angle, the fact that the state of an atom is expressed in terms of wave functions means that within it linear relationships in space and causal relationships in time stand in a close relationship to each other. The two types of relationship between our experiences that were produced in different forms by common sense were first unified by the principle of relativity into a four-dimensional world, a framework that forms a common background for time and space; then with quantum mechanics the relationship between the two was unified in still more concrete terms by means of wave functions. In a sense, the laws of modern physics are things that, in this broad sense, lay down simultaneously both the prevailing order and the development of nature. This, however, holds true only in a world of possibilities that, in a sense, transcends the world of facts.

If one turns back again, however, we find that what our common sense had taught us to think of as time was not something that, as in the theory of relativity, can be reduced to one more dimension added to the dimensions of space; it had its own independent direction. We had believed that the past and the future were strikingly different in nature. Yet once in the world of classical mechanics—and even more so if one took the theory of relativity into consideration—the one-directional nature of time was lost sight of. In quantum mechanics likewise, there is not, inherently, any intuitively discoverable obvious distinction between the past and the future in the world of possibilities that is expressed by wave functions. It is only when we come to the point where the world of possibilities and the world of facts relate to each other that we rediscover the lost properties of time. The past, in short, is an accumulation of facts. It is related to the actual world by our memories and records, as well as by heredity and relics of the past. The future lies before us as a world of possibilities. Through the workings of nature and man, some parts of it will be realized as facts, and at the same time will form new additions to the accumulation of the past. As science advances and our knowledge of the history of nature and man becomes richer and more accurate, we shall come to understand the past much more clearly than we do now. In actual practice, the shortage of information will probably always mean that the parts of the past that are still not clear will be far more numerous than those that are, but there are no limits in principle. On the other hand, though, the highest form of knowledge we can have concerning the future is the precise shape of the wave functions that determine the probability of various events occurring. In other words there is, generally speaking, no "law" that determines which of the innumerable facts awaiting realization in the future will be selected. The reason for this is that in cases where we are obliged to consider microscopic processes as the cause of a fact, causality is obliged to yield pride of place to the laws of probability that are more general than itself. For us, indeed, this may be good tidings rather than otherwise, for not only can we thereby transcend the world of actuality and enter the world of eternal truths, but we can also live in the world of actuality with ideals and with hope for the future.

From yet another angle, it may be said that the two different types of prevailing order that were distinguished by the common sense approach—the temporal and spatial relationships between concrete experiences, and the logical and mathematical relationships between abstract concepts and quantities—are related to each other in quantum mechanics in the form of actualities and probabilities. There is nothing odd in the fact that the word accurately expressing this relationship should have been "probabilities." It resembles the fact that the only accurate answer to the difficult question that worried the Greeks as to how many grains of wheat make a heap can be obtained by considering the question in terms of statistics or probability.

This account of the relationship between facts and laws has concentrated chiefly on those aspects affecting the exact sciences in general and modern physics in particular. In the other fields of natural science, and still more in the humanities, it may be necessary to interpret empirical fact in a wider sense. When it comes to questions of life, or of the relationship between matter and mind, a different way of thinking may be necessary. But where these questions are concerned, there would seem justification for believing that modern physics will be able to provide some new hints.

2 Meson Theory in Its Developments
Nobel Lecture, [1949]

The meson theory started from the extension of the concept of the field of force so as to include the nuclear forces in addition to the gravitational and electromagnetic forces. The necessity of introduction of specific nuclear forces, which could not be reduced to electromagnetic interactions between charged particles, was realized soon after the discovery of the neutron, which was to be bound strongly to the protons and other neutrons in the atomic nucleus. As pointed out by Wigner[1] specific nuclear forces between two nucleons, each of which can be either in the neutron state or the proton state, must have a very short range of the order of 10^{-13} cm, in order to account for the rapid increase of the binding energy from the deuteron to the alpha particle. The binding energies of nuclei heavier than the alpha particle do not increase as rapidly as if they were proportional to the square of the mass number A, i.e., the number of nucleons in each nucleus, but they are in fact approximately proportional to A. This indicates that nuclear forces are saturated for some reason. Heisenberg[2] suggested that this could be accounted for, if we assumed a force between a neutron and a proton, for instance, due to the exchange of the electron or, more generally, due to the exchange of the electric charge, as in the case of the chemical bond between a hydrogen atom and a proton. Soon afterwards, Fermi[3] developed a theory of beta decay based on the hypothesis by Pauli, according to which a neutron, for instance, could decay into a proton, an electron, and a neutrino, which was supposed to be a very penetrating neutral particle with a very small mass.

This gave rise, in turn, to the expectation that nuclear forces could be reduced to the exchange of a pair of an electron and a neutrino between two nucleons, just as electromagnetic forces were regarded as due to the exchange of photons between charged particles. It turned out, however, that the nuclear force thus obtained was much too small,[4] because the beta decay was a very slow process compared with the supposed rapid exchange of the electric charge responsible for the actual nuclear forces. The idea

of the meson field was introduced in 1935 in order to make up this gap.[5] Original assumptions of the meson theory were as follows.
I. The nuclear forces are described by a scalar field U, which satisfies the wave equation

$$\left(\frac{\partial^2}{\partial x^2} + \frac{\partial^2}{\partial y^2} + \frac{\partial^2}{\partial z^2} - \frac{1}{c^2}\frac{\partial^2}{\partial t^2} - \kappa^2\right) U = 0 \qquad (1)$$

in vacuum, where κ is a constant with the dimension of reciprocal length. Thus, the static potential between two nucleons at a distance r is proportional to $\exp(-\kappa r)/r$, the range of forces being given by $1/\kappa$.
II. According to the general principle of quantum theory, the field U is inevitably accompanied by new particles or quanta, which have the mass

$$\mu = \frac{\kappa\hbar}{c} \qquad (2)$$

and the spin 0, obeying Bose-Einstein statistics. The mass of these particles can be inferred from the range of nuclear forces. If we assume, for instance, $\kappa = 5 \times 10^{12}$ cm^{-1}, we obtain $\mu \cong 200\, m_e$, where m_e is the mass of the electron.
III. In order to obtain exchange forces, we must assume that these mesons have the electric charge $+ e$ or $- e$, and that a positive (negative) meson is emitted (absorbed) when the nucleon jumps from the proton state to the neutron state, whereas a negative (positive) meson is emitted (absorbed) when the nucleon jumps from the neutron to the proton. Thus a neutron and a proton can interact with each other by exchanging mesons just as two charged particles interact by exchanging photons. In fact, we obtain an exchange force of Heisenberg type between the neutron and the proton of the correct magnitude, if we assume that the coupling constant g between the nucleon and the meson field, which has the same dimension as the elementary charge e, is a few times larger than e.

However, the above simple theory was incomplete in various respects. For one thing, the exchange force thus obtained was repulsive for triplet S-state of the deuteron in contradiction to the experiment, and moreover we could not deduce the exchange force of Majorana type, which was necessary in order to account for the saturation of nuclear forces just at the alpha particle. In order to remove these defects, more general types of meson fields including vector, pseudoscalar, and pseudovector fields in addition to the scalar fields, were considered by various authors.[6] In particular, the vector field was investigated in detail, because it could give a combination of exchange forces of Heisenberg and Majorana types with correct signs and could further account for the anomalous magnetic moments of the neutron and the proton qualitatively. Furthermore, the vector theory predicted the existence of noncentral forces between a neutron and a proton, so that the deuteron might have the electric quadripole moment. However, the actual electric quadripole moment turned out to be positive in sign, whereas the vector theory anticipated the sign to be negative. The only meson field, which gives the correct signs both for nuclear forces and for the electric quadripole moment of the deuteron, was the pseudoscalar field.[7] There was, however, another feature of nuclear forces, which was to be accounted for as a consequence of the meson theory. Namely, the results of experiments on the scattering of protons by protons indicated that the type and magnitude of interaction between two protons were, at least approximately, the same as those between a neutron and a proton, apart from the Coulomb force. Now the interaction between two protons or two neutrons was obtained only if we took into account the terms proportional to g^4, whereas that between a neutron and a proton was proportional to g^2, as long as we were considering charged mesons alone. Thus it seemed necessary to assume further:

IV. In addition to charged mesons, there are neutral mesons with the mass either exactly or approximately equal to that of charged mesons. They must also have the integer spin, obey Bose–Einstein statistics and interact with nucleons as strongly as charged mesons.

This assumption obviously increased the number of arbitrary constants in meson theory, which could be so adjusted as to agree with a variety of experimental facts. These experimental facts could not be restricted to those of nuclear physics in the narrow sense, but was to include those related to cosmic rays, because we expected that mesons could be created and annihilated due to the interaction of cosmic ray particles with energies much larger than μc^2 with matter. In fact, the discovery of particles of intermediate mass in cosmic rays in 1937[8] was a great encouragement to further developments of meson theory. At that time, we came naturally to the conclusion that the mesons which constituted the main part of the hard component of cosmic rays at sea level was to be identified with the mesons which were responsible for nuclear force.[9] Indeed, cosmic ray mesons had the mass around 200 m_e as predicted and, moreover, there was the definite evidence for the spontaneous decay, which was the consequence of the following assumption of the original meson theory:

V. Mesons interact also with light particles, i.e., electrons and neutrinos, just as they interact with nucleons, the only difference being the smallness of the coupling constant g' in this case compared with g. Thus a positive (negative) meson can change spontaneously into a positive (negative) electron and a neutrino, as pointed out first by Bhabha.[10] The proper lifetime, i.e., the mean lifetime at rest, of the charged scalar meson, for example, is given by

$$\tau_0 = 2 \left(\frac{\hbar c}{(g')^2} \right) \left(\frac{\hbar}{\mu c^2} \right) \tag{3}$$

For the meson moving with velocity v, the lifetime increases by a factor $1/\sqrt{1-(v/c)^2}$ due to the well-known relativistic delay of the moving clock. Although the spontaneous decay and the velocity dependence of the lifetime of cosmic ray mesons were remarkably confirmed by various experiments,[11] there was an undeniable discrepancy between theoretical and experimental values for the lifetime. The original intention of meson theory was to account for the beta decay by combining the assumptions III and V together. However, the coupling constant g', which was so adjusted as to give the correct result for the beta decay, turned out to be too large in that it gave the lifetime τ_0 of mesons of the order of 10^{-8} sec, which was much smaller than the observed lifetime 2×10^{-6} sec. Moreover, there were indications, which were by no means in favor of the expectation that cosmic-ray mesons interacted strongly with nucleons. For example, the observed cross section of scattering of cosmic-ray mesons by nuclei was much smaller than that obtained theoretically. Thus, already in 1941, the identification of the cosmic-ray meson with the meson, which was supposed to be responsible for nuclear forces, became doubtful. In fact, Tanikawa and Sakata[12] proposed in 1942 a new hypothesis as follows: The mesons which constitute the hard component of cosmic rays at sea level are not directly connected with nuclear forces, but are produced by the decay of heavier mesons which interacted strongly with nucleons.

However, we had to wait for a few years before this two-meson hypothesis was confirmed, until 1947, when two very important facts were discovered. First, it was discovered by Italian physicists[13] that the negative mesons in cosmic rays, which were captured by lighter atoms, did not disappear instantly, but very often decayed into electrons in a mean time interval of the order of 10^{-6} sec. This could be understood only if we supposed that ordinary mesons in cosmic rays interacted very weakly with nucleons. Soon afterwards, Powell and others[14] discovered two types of mesons in cosmic rays, the heavier mesons decaying in a very short time into lighter mesons. Just before the latter discovery, the two-meson hypothesis was proposed by Marshak and Bethe[15] independent of the Japanese physicists above mentioned. In 1948, mesons were created artificially in Berkeley[16] and subsequent experiments confirmed the general picture of two-meson theory. The fundamental assumptions are now:[17]

(i) The heavier mesons, i.e., π-mesons with the mass m_π about 280 m_e interact strongly with nucleons and can decay into lighter mesons, i.e., μ-mesons and neutrinos with a lifetime of the order of 10^{-8} sec; π-mesons have integer spin (very probably spin 0) and obey Bose-Einstein statistics. They are responsible for, at least, a part of nuclear forces. In fact, the shape of nuclear potential at a distance of the order of $\hbar/m_\pi c$ or larger could be accounted for as due to the exchange of π-mesons between nucleons.

(ii) The lighter mesons, i.e., μ-mesons with the mass about 210 m_e are the main constituent of the hard component of cosmic rays at sea level and can decay into electrons and neutrinos with the lifetime 2×10^{-6} sec. They have very probably spin $\frac{1}{2}$ and obey Fermi-Dirac statistics. As they interact only weakly with nucleons, they have nothing to do with nuclear forces.

Now, if we accept the view that π-mesons are the mesons that have been anticipated from the beginning, then we may expect the existence of neutral π-mesons in addition to charged π-mesons. Such neutral mesons, which have integer spin and interact as strongly as charged mesons with nucleons, must be very unstable, because each of them can decay into two or three photons.[18] In particular, a neutral meson with spin 0 can decay into two photons and the lifetime is of the order of 10^{-14} sec or even less than that. Very recently, it became clear that some of the experimental results obtained in Berkeley could be accounted for consistently by considering that, in addition to charged π-mesons, neutral π-mesons with the mass approximately equal to that of charged π-mesons were created by collisions of high-energy protons with atomic nuclei and that each of these neutral mesons decayed into two mesons with the lifetime of the order of 10^{-13} sec or less.[19] Thus, the neutral mesons must have spin 0.

In this way, meson theory has changed a great deal during these fifteen years. Nevertheless, there remain still many questions unanswered. Among other things, we know very little about mesons heavier than π-mesons. We do not know yet whether some of the heavier mesons are responsible for nuclear forces at very short distances. The present form of meson theory is not free from the divergence difficulties, although recent development of relativistic field theory has succeeded in removing some of them. We do not yet know whether the remaining divergence difficulties are due to our ignorance of the structure of elementary particles themselves.[20] We shall probably have to go through another change of the theory, before we shall be able to arrive at the complete understanding of the nuclear structure and of various phenomena, which will occur in high energy regions.

Notes

1. E. Wigner, *Phys. Rev.*, 43 (1933) 252.
2. W. Heisenberg, *Z. Physik*, 77 (1932) 1; 78 (1932) 156; 80 (1933) 587.
3. E. Fermi, *Z. Physik*, 88 (1934) 161.
4. I. Tamm, *Nature*, 133 (1934) 981; D. Ivanenko, *Nature*, 133 (1934) 981.
5. H. Yukawa, *Proc. Phys.-Math. Soc. Japan*, 17 (1935) 48; H. Yukawa and S. Sakata, ibid., 19 (1937) 1084.
6. N. Kemmer, *Proc. Roy. Soc. London*, A 166 (1938) 127; H. Fröhlich, W. Heitler and N. Kemmer, ibid., 166 (1938) 154; H. J. Bhabha, ibid., 166 (1938) 501; E. C. G. Stueckelberg, *Helv. Phys. Acta*, 11 (1938) 299; H. Yukawa, S. Sakata, and M. Taketani, *Proc. Phys.-Math. Soc. Japan*, 20 (1938) 319; H. Yukawa, S. Sakata, M. Kobayashi, and M. Taketani, ibid., 20 (1938) 720.
7. W. Rarita and J. Schwinger, *Phys. Rev.*, 59 (1941) 436, 556.
8. C. D. Anderson and S. H. Neddermeyer, *Phys. Rev.*, 51 (1937) 884; J. C. Street and E. C. Stevenson, ibid., 51 (1937) 1005; Y. Nishina, M. Takeuchi, and T. Ichimiya, ibid., 52 (1937) 1193.
9. H. Yukawa, *Proc. Phys.-Math. Soc. Japan*, 19 (1937) 712; J. R. Oppenheimer and R. Serber, *Phys. Rev.*, 51 (1937) 1113; E. C. G. Stueckelberg, ibid., 52 (1937) 41.
10. H. J. Bhabha, *Nature*, 141 (1938) 117.

11. H. Euler and W. Heisenberg, *Ergeb. Exakt Naturw.*, 17 (1938) 1; P. M. S. Blackett, *Nature*, 142 (1938) 992; B. Rossi, *Nature*, 142 (1938) 993; P. Ehrenfest, Jr. and A. Fréon, *Compt. Rend.*, 207 (1938) 853; E. J. Williams and G. E. Roberts, *Nature*, 145 (1940) 102.
12. Y. Tanikawa, *Progr. Theoret. Phys. Kyoto*, 2 (1947) 220; S. Sakata and K. Inouye, ibid., 1 (1946) 143.
13. M. Conversi, E. Pancini, and O. Piccioni, *Phys. Rev.*, 71 (1947) 209.
14. C. M. G. Lattes, H. Muirhead, G. P. S. Occhialini, and C. F. Powell, *Nature*, 159 (1947) 694; C. M. G. Lattes, G. P. S. Occhialini, and C. F. Powell, *Nature* 160 (1947) 453, 486.
15. R. E. Marshak and H. A. Bethe, *Phys Rev.*, 72 (1947) 506.
16. E. Gardner and C. M. G. Lattes, *Science*, 107 (1948) 270; W. H. Barkas, E. Gardner, and C. M. G. Lattes, *Phys. Rev.*, 74 (1948) 1558.
17. For further details, see H. Yukawa, *Rev. Mod. Phys.*, 21 (1949) 474.
18. S. Sakata and Y. Tanikawa, *Phys. Rev.*, 57 (1940) 548; R. J. Finkelstein, ibid., 72 (1947) 415.
19. H. F. York, B. J. Moyer, and R. Bjorklund, *Phys. Rev.*, 76 (1949) 187.
20. H. Yukawa, *Phys. Rev.*, 77 (1950) 219.

3 Space-Time and Elementary Particles [1963]

The Discovery of Elementary Particles

Elementary particles have been generally understood to be the smallest individual units of matter, typically possessing both wave-like and particle-like properties. Initially it seemed that the total number of such particles was very small, namely the electron, the proton, and the photon (with zero rest mass). The present situation is very different, for the total number of kinds of such particles now far exceeds the number of kinds of chemical elements. This is not a very appealing concept, and it is difficult to avoid the suspicion that a simpler fundamental reality lies behind these particles, in terms of which they can be explained.

The early history of this subject dates from the end of the

nineteenth century when the electron was discovered and recognized as the first of what we now call the elementary particles.
Soon after the turn of the century, light was also recognized as
possessing particle-like properties, according to quantum theory.
Matter and energy could thus both be considered as particle-like
manifestations of nature. On the other hand the theory of relativity revealed the equivalence of mass and energy. This gave rise
to a new synthesis, in which the elementary particle assumed the
role previously played by the atom.

The discovery of the phenomenon of radioactivity at about the
same time as that of the electron also played its part. This extraordinary phenomenon, with the emission of prodigious amounts
of energy from the atom, was quite without parallel and provided
important clues to the structure of matter. It led eventually to the
discovery in the early years of the twentieth century that the atom
consists of both electrons and a nucleus and led further to the
achievement of the alchemist's dream, the transmutation of the
elements.

With these and several other momentous discoveries, most
notably that of the neutron in 1932, the world of the elementary
particle has gradually begun to emerge.

The term "elementary particle" itself has had a long history,
but for much of that time was without any clearly defined meaning. The proton (the nucleus of the hydrogen atom) was early
considered to be an elementary particle, as was the neutron upon
its subsequent discovery. Eventually the proton, neutron, electron
and photon were accepted as the basic constituents of nature.

The year 1932 also yielded the discovery of the positron, a
particle which although very closely associated with the electron
differs from it in carrying a positive charge, whose existence was
already predicted by Dirac. Pauli had already in the previous year
suggested the existence of a another kind of particle, the neutrino.
This view was not printed in any of the scientific journals, so that

it was not until a few years later that I became aware of it. The direct experimental proof of the existence of the neutrino came much later.

In 1935 the meson also appeared on the scene. Thus the number of kinds of particles had considerably increased, but did not appear unreasonably large, and the existence of each had its own reason. Such was the situation up until the forties, when the two-meson theory was propounded by Sakata and Tanikawa. The question immediately arose as to why nature needs two kinds of meson; it was felt that one of them ought to be redundant. However, confirmation of the existence of both kinds of meson in cosmic rays came in 1947. Indeed, cosmic rays have been acting like a gigantic natural accelerator in providing a source of hitherto unidentified particles.

Even stranger particles were also found in the cosmic rays in 1947, at that time called simply "V-" or "new" particles. Experimental investigations of cosmic rays revealed that there was a multiplicity of these particles too, a quite unprecedented situation. It was just because elementary particles had been regarded as the ultimate units of matter and energy, and therefore of the whole physical world, that the appearance of so many kinds of them posed a difficult new problem.

With the development of large-scale particle accelerators from 1947 onwards it became possible to generate mesons artificially. Subsequent increases in the size and energy of accelerators eventually enabled the generation of the new or V-particles and many other mysterious kinds of particles.

The Concept of a Composite Particle

In the fifties a total of about thirty different kinds of particle were discovered, a very different situation from that of the thirties. Furthermore, in the sixties a new generation of particle accelerators with enhanced capabilities had enabled greater precision in experimental results, so leading to the discovery of numerous particles with extremely short lifetimes. It is not clear whether these should be classified as elementary particles. If we should simply classify them as elementary particles without qualification, then the total number runs into the hundreds, which poses considerable conceptual difficulties. It seems more reasonable under these circumstances to abandon the concept that what we have called elementary particles are in fact an ultimate unit.

When we get down to details, views on these points tend to diverge, but the simplest and the one which is a simple extension of the course of historical development is as follows.

The ultimate unit was originally thought to be the atom, but this was later found to be composed of electrons and a nucleus. Further analysis revealed that still more fundamental units existed, through the discovery that the nucleus consists of protons and neutrons. It follows that the same may well hold for the elementary particle; most of them would then not be fundamental, but would in fact be made up of a very few kinds of really fundamental particles. Sakata's theory of composite particles is particularly worthy of attention among subsequent developments of this view.

Sakata's original concept was that the fundamental particles would be only three in number, namely the proton, the neutron, and the lambda particle V, which is one of the V-particles previously mentioned. There seems no special objection to the number 3, but in other respects there are serious difficulties. Probably we would look for some other fundamental triplet. However, if one assumes that there is only one set of fundamental triplets, which is called "quarks" by Gell-Mann, it would have to possess some very odd properties. The existence of such a triplet has yet to be verified experimentally. We must admit the possibility that it might not be simply particles, but some new physical entity not identifiable as particles in the usual sense. This means more or less a departure from the composite particle theory.

We should perhaps mention here a peculiar group of particles formed by the electron, the neutrino, and the muon (one form of meson). This group makes up a special family of particles called leptons. One view has it that these are the fundamental particles, with the neutrino possibly being the absolutely fundamental one. The idea originated with the French scholar de Broglie, who suggested that the photon in fact consists of two neutrinos. This concept has undergone various modifications in later developments, and Taketani has adopted a similar view of neutrino monism. This is rather akin to the idea mentioned previously, but does differ in some important respects.

The Bootstrap Theory

Heisenberg expressed the view that we may need to consider nothing more abstruse than elementary particles as we know them, nor look for some unfamiliar manifestation of particles. Considering the train of thought we have developed so far, from molecule to atom, from the atom to the nucleus, and from the nucleus to the proton and the neutron; in each of these instances a molecule, say, actually breaks up to form atoms, and atoms do in fact combine to form molecules. Further, the nucleus undoubtedly lies within the atom, and disintegration of a nucleus yields protons and neutrons. Conversely, recombination of individual protons and neutrons to form a nucleus is also possible. The disintegration into separate units and recombination into structurally more complicated assemblages not only occurs reversibly in nature, but may also be duplicated artificially. This is not found to be so when we come to the case of the elementary particles, and the same ways of thinking are no longer applicable.

As an example we may take the case of a neutron, which meson theory tells us will become a proton upon the addition of a positively charged π-meson, and the reverse is also true. However, it would be wrong to conclude from this that the proton is a composite particle consisting simply of a neutron and a positively charged π-meson. There is equal justification for regarding the neutron as a composite particle consisting of a proton and a negatively charged π-meson.

It is also impossible to establish any kind of hierarchy of essential simplicity among the great variety of new particles that have been observed with the increase in energy of particle accelerators. In particular, even if a given particle decays into two others, it is impossible to say that the fragments are either simpler than the original particle or really fundamental. Furthermore, the mode of decay is not necessarily unique, and indeed when a particle is subjected to bombardment by other particles the resultant disintegration products may well be more complex particles rather than less.

Heisenberg therefore urges that rather than seeking to classify any of these particles as fundamental, we should be content with the observation that they exhibit transmutability among themselves.

Several scholars in the United States share this idea. For instance the theory of Chew suggests that all elementary particles are composites of at least two other elementary particles, and that their relationship is reversible in that no one of them is dominant over the others. Namely any one elementary particle can serve as an element of various others. He calls this the "bootstrap" relationship.

One may say that this is a kind of "natural metempsychosis." Although such a conception is certainly interesting, I think we would run into considerable difficulties trying to define the term precisely. There is also the question of whether the mere introduction of such a term would in fact solve our problems. Heisenberg seeks to educe the origins of elementary particles from a still more fundamental *Urmaterie* ("primordial matter"), rather than deal with the particles themselves. The aim is to educe all the phenomena associated with fundamental particles from the theory of *Urmaterie* all at once. In essence it is a variant of the unified theory, a feature which Heisenberg's monistic concept, the composite particle theory, and Chew's idea of the bootstrap all have in common.

Since it seems as if we are unlikely to proceed very far by concentrating on individual particles, we certainly need a theory which can give us an overall perspective of the whole range of elementary particles. In this connection, I will now give some consideration to the development of my own views on the same problem.

Nonlocal Field Theory

The year 1942, I think, marked a watershed in theoretical research into elementary particles in Japan. It was in that year that Sakata and Tanikawa presented their two-meson theory. Related ideas also lie behind Sakata's theory of composite particles presented later in the fifties.

However, another idea was evolving in the same year. That is to say, I myself was considering assigning a physical extension to the elementary particle. If the elementary particle is considered to be localized at a point, having zero volume, this affects the field about the point. For example in the case of an electron, if the electric charge is assumed to be concentrated at a point, then the electric field in the immediate vicinity will be very intense. Calculations of the energy of such a field give values which diverge to infinity. This is no new problem, but just the simplest manifestation of the problem of "divergence" or "infinite singularities," a problem which is not limited to quantum mechanics, but is also encountered in the earlier classical electrodynamics.

A physical theory must be considered fundamentally defective if it predicts that observable quantities should be infinitely large, but it is a defect exhibited by the theory of relativity, and which persists in quantum mechanics as long as we take special relativity seriously. The form in which we encounter the difficulty has undergone gradual change, but we are forced to admit that it has not yet been resolved. My contention was that we ought to set a high priority to assigning some kind of spatial extension to primary particles, and work along this line has continued up to the present time.

Yet another view was put forward by Tomonaga, who, being reluctant to assign any extension to the particles even after listening to my repeated talks, attempted to transform the mathematical formulations of quantum field theory so that they were in closer accord with the theory of relativity. This is known as the "super-many-time" theory, and I felt it very interesting. Employing this concept, the approach to the problems of singularities or divergences became somewhat different. The divergences remain, but can be rendered innocuous. That is to say that when handled correctly they do not prevent quantum electrodynamics from giving results in surprisingly good agreement with experiments.

Nevertheless I have myself by no means abandoned the idea of assigning an extension to the elementary particle. By 1950 or a little earlier I had put my ideas into the form of a thesis, now known as the theory of nonlocal fields, where I refer to a four-dimensional extension.

Classification of the Elementary Particles

If we delve a little deeper into the origin of these concepts, I think it was in the thirties that Markov of the Soviet Union first proposed that in quantum electrodynamics the electromagnetic field should be considered in terms we now describe as nonlocal, but the idea did not undergo very much development at that time, By around 1950, when I actually presented my theory of nonlocal fields, various kinds of new particles had already emerged. This gave me the additional idea that as well as visualizing the particles as having an extension it might be legitimate to assume that this gives each of the particles some inner structure, which could help to explain the variations in their properties.

In the meantime the number of different kinds of particles increased to such an extent that it enabled us to classify and group them. In the fifties Nishijima, Nakano, and Gell-Mann simultaneously presented papers which were very useful in the classification of heavier particles and the mesons, though not for photons and the leptons. Their classifications involved the use of various quantum numbers, and amounted to a series of statements that a particular particle had quantum numbers which took these values, while another had different values, etc.

Even today, although we know the effectiveness of this system, we cannot attach any physical meaning to the newly appearing quantum numbers. Of those old quantum numbers which are understood, the most typical is probably the spin quantum number. The classical analogue of spin is exhibited in the diurnal rotation of the earth, whether it be clockwise or anticlockwise, fast or slow, in whatever direction the axis is pointing, and whatever the total angular momentum might be. In the normal world accessible to our senses, spin can take any arbitrary value within a continuous range of directions and magnitudes. In the world of the elementary particles like the electron, we have discovered that spin can only take certain values, and that there is a basic unit of spin. Measuring with this unit we find that only certain discrete multiples are ever observed, that is zero, and integral and half-integral numbers. It seems very strange that continuous variation is not possible, but in the world of quantum mechanics this is undoubtedly so. Nevertheless, quantum spin is a concept we grasp through its associations with rotation. Since it is related to something which is familiar from the everyday world, we can still understand the concept however far the process of abstraction may have been taken.

When we come to new quantum numbers like isospin, strangeness, and others which spring from the attempt to classify all particles including the new, although they apparently refer to the behavior of the particle, we cannot in fact make any association with something in our space-time world. On the other hand, when we consider nonlocal fields, with the elementary particle no longer regarded as a point but having a certain extension, then something like spin can be associated with relative motion of the different parts of the extension. Other quantum numbers may also be derived, and could in general be associated with some difference in the internal motion in our space-time world.

This approach would at least give us some help in grappling with the problem in our attempt to understand it. The theory of composite particles is a similar attempt to reach an understanding of the question, but the difference is whether there are really fundamental particles which are again point particles described by local fields, or not.

The Nature of the Four-Dimensional Extension

Continuing our discussion of nonlocal fields a little further, we know that an "extension" is assigned to the particle, but there can be several different kinds of extension. A molecule, for instance, may consist of two atoms. Two hydrogen atoms come together to form a hydrogen molecule. Within the molecule, then, there are two nuclei, that is two protons, and they are in relative motion, revolving and vibrating. My first conception of the simplest kind of nonlocal field was not very different from this.

Let us consider hydrogen molecules further. One possibility is to regard them as two-point particles with a certain complex force acting between them. Under the action of this force they vibrate and rotate. The relevant concept in the theory of nonlocal fields is more subtle. The particles associated with the nonlocal field are not considered as an assembly of two-point particles, but there are still several features of the analysis which show strong resemblances to such a system. The subsequent discovery of so many new particles meant that it was no longer sufficient even to consider two points rather than one in a nonlocal field. We eventually came to think of nonlocal fields in a much broader sense, as embracing three or four points simultaneously.

I worked on this theory with Katayama of Kyoto University, while Takabayashi of Nagoya University was working on a similar idea. There is, however, yet another method of approach. It can be classified with the general category of nonlocal field theories and employs the conception of extension, but it is somewhat simpler. It differs in that the extension is assumed to be both initially and subsequently fully occupied by something.

The simplest analogy is probably with a rigid body, which can both rotate and move as a whole. No actual substance fully satisfies the conditions of complete rigidity but there are many solid bodies which approximate to them very closely. Our concept of a rigid body is an idealization based on phenomena which are accessible to us. We may regard elementary particles as a kind of rigid body. Naturally, we must think of them as relativistic rigid bodies in the four-dimensional world.

In the fifties Nakano worked on the implications of this, also followed up by de Broglie, Vigier, and others in France. Hara of Nihon University also continued to work along this line, and in the sixties he investigated the effect of permitting slight deformations, treating them quantum-mechanically and trying to make them consistent with the theory of relativity too. This is a rather technical point, but the basic idea is still simply that the extension is fully occupied by something like a quantized ether.

It should be noted that we are not here at all concerned with the kind of discontinuity which was revealed as point particles in vacuum. Our concern is rather the nature of the extensions of the basic particles themselves, the protons, neutrons, electrons, and others.

To conclude this section, let us return for a moment to the example of spin. The quantum number can assume only discrete values, that is zero or integers and half integers. It is extremely difficult to understand why it should be able to assume the half-integral values. Suppose we have two-point particles in relative motion. If we assume that one is very heavy and at rest, then the other may be supposed to revolve around it. The spin associated with such an orbital motion must be either zero or take some integral value. It is impossible to associate a half-integral spin with the motion defined by these conditions.

Despite the fact that the electron was assumed to be confined to a point, Dirac succeeded in deriving a spin proper to the electron itself quite apart from its orbital motion based on a relativistic treatment of the motion; in doing so he was able to account for half-integral spins. We can derive the same result as a consequence of assigning an extension to the particle. It is an encouragement to know that if a rigid body is formed from a material continuum, the spin can take half-integral as well as integral values, whereas if it is held to consist of an assemblage of point particles this is not possible, unless certain constraints on the geometrical configuration of the assemblage are imposed. It would therefore seem reasonable to work from the outset in terms of a continuum rather than a system of point particles.

"The universe, a wayside inn for all things. . ."

In speaking of a body made up of a material continuum we usually have in mind something with a definite size and shape, and which is fully occupied by the constituent material. The difficulty arises when we try to make our conception consistent with the requirements of quantum mechanics and relativity theory. Perhaps we can approach the problem by first considering an energy-free void. That is to say, we isolate a definite elemental energy-free domain of our world of space and time.

If energy in any form comes to be associated with the void, then dependent upon the manner of that association we can regard it as a matter- or particle-like manifestation, even an elementary particle. If we envisage our domain becoming infinitesimally small, it would in the limit be equivalent to a point particle, and our theoretical formulations would encounter the same difficulties as before. We therefore set some lower limit to the size of such a domain, a limit which corresponds to the smallest quantum of space-time, that is a domain which cannot be meaningfully further subdivided. This we may call an elementary domain. If energy is associated with each elementary domain, dependent upon the mode of association, the domain would be identifiable as a variety of different elementary particles. But the domain itself is there, even if it is empty.

The concept of empty domain is, for myself, an expression of the strong formative influence that the philosophy of Laotse and Chuangtse has had on my ideas. There is a phrase which appears at the beginning of one of the works of Li Po, the great Chinese poet of the T'ang dynasty, who was a follower of Laotse and Chuangtse, "The universe is a wayside inn for all things . . ." followed by the phrase, "Time is an eternal wayfarer." This appealed so much to the famous poet Basho that he opened his book *Oku no Hosomichi* with it.

Turning from Basho and Li Po, and seeking a modern interpretation for the statement that "the universe is a wayside inn for all things," we may liken the four-dimensional space-time continuum to an inn which provides accommodation for the whole of creation. Such a conception, perhaps, retains something of the spirit of the philosophy of Laotse and Chuangtse.

In the discussion developed so far, when I have referred to "relativity theory" it has been to the "special theory of relativity." We must not fail, however, to recognize the difference between it and the general theory of relativity. The latter, which is concerned with universal gravitation, is uniquely different from all other physical theories and is particularly well suited to discussions of the universe as a whole. Geometrical characteristics of the four-dimensional space-time continuum are shown to be operative within the world of physics in a quite unexpected way. No one other than Einstein could have expected that universal gravitation, on the face of it a purely physical concept, could be identified with certain geometrical properties of the space–time continuum. Einstein not only established this through his development of the general theory of relativity, but in doing so succeeded, to an extent that I regard as no less than miraculous, in achieving a more powerful and radical synthesis in his theory of gravitation than Newton had done. My own development of the concept of elementary domains owes much to Einstein's thought in this respect.

Einstein visualized the macroscopic world in terms of a four-dimensional geometry on an immensely large scale. The microscopic space-time world presents greater conceptual difficulties, and very few workers have so far grappled with its problems. In the special theory of relativity, four-dimensional Minkowski space merely provides a frame of reference, having a fixed geometrical structure from the outset, and is assumed to retain this fixed structure whatever may happen in this world. Our conceptions of matter and energy are set within this fixed frame of reference. However, the question remains: should we not also examine the relationship between space and time in the subatomic world, that is the world of the elementary particles in much the same spirit as that of general relativity?

In introducing elementary domains into the discussion we are really assigning limits to the extent to which time and space can be subdivided.

Metempsychosis or Transmigration

Returning once again to the ideas of Chuangtse, "chaos" in his writings is very much akin to the world of the elementary particle. He says that to attempt unskillfully to impose some kind of physiognomy upon this chaos would be to destroy it. Although this kind of statement will mean different things to different people, it seems to me to be hinting at the kind of situation we meet in the context of elementary particles. I do not wish to overplay the point, but it does seem that modern physics in many respects carries an echo from ancient philosophies. Indeed, we do sometimes find that a flash of inspiration in the mind of a scholar of the ancient world sheds a startling ray of illumination upon a discovery made much later. This is why the "atom" of Democritus is so astonishing. We have something quite miraculous in the contributions to the world of thought from Greece, Israel, China, and India.

Plato attributes to Socrates the statement that man gains knowledge by recollection. The key thought here was probably that of reincarnation, or the transmigration of the soul. What happened to Democritus's thought from the remote past, however, is more like resurrection! Newton seemed to have in his mind a kind of atomic philosophy, but it was only as late as the nineteenth century that Dalton presented a new image which was clearly similar to that of Democritus. May we not regard this as a kind of "recollection"?

In the case of my own experience, although having long forgotten my original conception of 1942, I feel that I have eventually recalled it in a different form. Careful reflection, however, shows my thought now to be unexpectedly similar to that of nearly thirty years ago. I may be able to go back further to those early days when I knew very little of physics, but I did have contact with ancient philosophers.

This brings us back to my previous comments on the empty domain. The nature of elementary particles cannot be considered in isolation from the structure of space itself. If the space cannot be subdivided without limit, the displacement from one place to another can also not be arbitrarily small. Thus, we may introduce new mathematical relations into the world of physics, whereby differential equations are replaced with "difference" equations.

Nature's Discontinuities

Newton and Leibniz were the first to develop the differential calculus, and Newton brought his theory of dynamics to completion by applying it to motion. Motion, in his sense, is completely continuous and can be divided into arbitrarily small segments. The position of a body undergoes an infinitesimal displacement in a correspondingly short length of time. This yields a definition of the velocity at a given moment and enables Newton's law of motion to be expressed in terms of differential equations. The motion can be determined when the equations are solved. The twentieth century, however, has seen the development of quantum mechanics, which deals with discontinuous as well as continuous processes. Discontinuity is inherent in the very nature of the atom—but we also encounter a discontinuity in the time-dependence of phenomena which is quite inconceivable in the ordinary world; nature actually jumps!

If we suppose that a man was initially at a certain place and that subsequently he is at some place remote from the former, we know that he must have traced a continuous path between them. However, in the quantum-mechanical world, energy sometimes jumps from one place to another and we cannot say that it has passed through intermediate points.

The elementary domain is not infinitesimally small. It has a certain extension, not only in the spatial dimensions but also in that of time. The fundamental equations that determine the properties of elementary domains must therefore do so in terms of finite increments of time, and similarly in spatial displacement. It is this idea which has absorbed my attention in the last few years.

Perception in the Subatomic World

When using difference equations, for instance if we are concerned with the change that has taken place after, say, one second, we do not concern ourselves with how much change would have taken place after only half a second nor do we ask about the change occurring in one tenth of a second. To take an example from our own experience, consider our reaction when we see a cup. We recognize in an instant that it is a cup. Actually our eyes trace momentarily the outline of the cup, and it is only after this admittedly very brief interval that we recognize the cup as a cup. This so-called pattern recognition takes a finite amount of time. We also have an afterimage which persists for some little time after we have seen the cup. We say that we recognize something in an "instant," but it is in fact a measurable period of time. We cannot say what has happened halfway through that period, and we must admit that recognition of the cup would become quite impossible if we think in terms of even smaller fractions of the period. It may be an instant to man, but on the subatomic scale an immense period of time elapses. Similarly if the cup moves we think of it as having moved continuously. This is certainly true, but if we analyse the process of its recognition in our body, we shall find it not so simple and smooth as the motion of the cup in the outside world.

The best example is an extremely common one. When faced with a differential equation which presents intractable problems, the next step is usually to try to resolve them with the help of a digital computer. This means solving the equation after restating it in the form of a difference equation. All the computer does is to perform the necessary arithmetical calculations instead of directly solving differential equations.

The human brain has truly marvellous capabilities, and geniuses like Newton and Leibniz have arisen, able to develop mathematical methods capable of dealing with continuous changes. However, a digital computer at the most can perform extremely complicated arithmetic at very high speed.

When our problems concern time dependence, instead of solving the differential equation, we work with very fine increments of time. In place of one second we choose one tenth of a second— and substitute one tenth with one hundredth of a second, and so on. In this way we present the computer with what is in effect a difference equation. If we can assume that whatever may happen within our smallest time interval (in our example one hundredth of a second) will not affect the situation at the end of the interval, then we may confidently predict the change after successive intervals and ignore any changes within the interval. To put this in another way, the differences that may arise there can all be tolerated. In solving a difference equation we generally derive far more solutions than we would from a differential equation. From among the multitude of solutions, if we could select those which are relevant and somehow systematize them, then we might be able to overcome the problem of infinity which has been so troublesome for so long. We have already been performing some of these very complex calculations, and the results have been promising.

The Long Stagnation of Research

One regrettable trend in theoretical physics over the last three decades has been to treat the various problems only within a preconceived framework.

Recapitulating briefly, the first thirty years or so of this century

184 • ON THEORETICAL PHYSICS

can be said to have been characterized by bold new ideas. The transition from atomic physics to nuclear physics and the realm of the elementary particles began in the thirties. Concerning the present state of progress in elementary particle physics, my own personal view is that it ought not to have stagnated for quite this long. The number of research workers has increased, and among them are men of great ability. The fault surely lies in an excessive conservatism in the realm of ideas, retaining the same preconceptions and pursuing the same lines of development—an unfortunate and inefficient process. Even while keeping to one particular line of development the objective will rarely be attained without some radical leap forward along the way. I myself may have been a little too hasty when I made that fateful leap long ago. It may have been foolhardy for a man of middle age like myself to go so far beyond the accepted bounds, but young people are supposed to have a spirit of adventure. I do hope that some really bold new ideas will come out sooner or later from among young physicists.

V   ON PEACE

I   Atomic Energy and
the Turning Point for Mankind [1954]

In primitive times, man succeeded in domesticating wild beasts. In the twentieth century, man has created a new and terrifying beast of his own. From the moment when scientists first realized the possibility of the use of atomic energy, it could be foreseen that it could become either a useful domestic animal or a wild beast of unparalleled ferocity. First, the atomic bomb showed the ferocity at its most naked. Large numbers of Japanese fell victim to it. Yet the fact that humanity as a whole was also the victim at that time has not, I feel, been fully appreciated in the world at large.

Those engaged in producing the atomic bomb had confidence that they were controlling the beast. To those who suffered by it, it was a ferocious brute and nothing else. To the owner, it may have appeared like a watchdog that faithfully performed its appointed task and no more. Yet the ferocity of atomic energy grew intenser day by day, and the destructive powers of the hydrogen bomb are greater, even, than those directly concerned had foreseen. Now once again, due to the test at Bikini, Japanese have suffered. There is also said to have been a considerable number of victims among the inhabitants of the islands close to the testing ground—which was something that even those carrying out the test had not expected. The beast that is atomic energy seems to be displaying a ferocity that even its owner cannot control. It was used in this case, moreover, not directly for purposes of war but in an experiment carried out to test the power of the bomb in an area considered comparatively safe. What is more, all kinds of devices are conceivable, even now, that would make the losses suffered by mankind still more widespread. The relationship between atomic energy and mankind has entered on a new and still more dangerous stage. The one ray of light in the gloom is that the recognition of an obvious fact—that the harm suffered by Japanese in the Bikini test is harm suffered by all mankind—is spreading throughout the world more rapidly and with a greater sense of actuality than in the case of Hiroshima and Nagasaki.

By now it has become clear beyond all doubt that the problem of atomic energy will be, for some time to come at least, the greatest problem facing mankind. It is essentially different from the problem faced by primitive man, that of how to protect himself from the wild beasts, for the threat posed by the power of the atom has its origins in knowledge that twentieth-century man has acquired for himself. It is possible to exterminate wild beasts, but not to eliminate scientific knowledge; for scientific knowledge can be stored in the human brain and shared among infinite numbers of men. Many conditions, of course, must be fulfilled before knowledge of the atom can enable us to bring atomic energy under control and so lead to the atomic bomb. Whatever form the study of atomic energy may take in contemporary Japan, for example, it should be quite unthinkable that Japan should within any short space of time produce an atomic bomb of her own. For some time to come, it seems, a small number of powerful nations will continue to maintain their positions as owners of the beast. The chief responsibility for handling the question of atomic energy obviously rests with these owners. Yet the question itself is not one for a small group of powerful states but for mankind as a whole. It is certain, too, that should the beast be turned into a useful domestic animal the whole of mankind would benefit greatly. The peaceful use of atomic energy is something capable of realization even in nations that are not so powerful in themselves. Already a considerable number of nations are advancing in this direction. While this is undoubtedly a bright source of hope for mankind, the ferocity of the wild beast is also being stepped up day by day. At the stage of the atomic bomb, the beast belonged to the same breed as the domestic animal that could work in the service of man; at the stage of the hydrogen bomb, it is a different breed that is far less susceptible to domestication. So far, at least, all attempts at domesticating it have failed.

The question of atomic energy is a question for all mankind. It has its origins, moreover, in the scientific knowledge stored in the minds of men. Any fundamental solution of the problem must likewise, it seems, emerge from men's own minds. It must begin with the awareness that a new question has appeared in the course of man's evolution, one that can determine his whole fate. For mankind to protect itself from the threat of atomic energy is, surely, a goal that should be given precedence over all others. The prosperity and happiness of mankind ought, properly speaking, to be common aims concerning which no dissidence is possible. In most cases, though, these have remained ideals far removed from the actual life of the individual. In practice, men have been moved by more pressing motives. Even though the ultimate objectives might be the prosperity and happiness of mankind, many differing views were possible in the past concerning the correct way to attain them. Religious inclination, for example, has often been a determining element in human group activity; in some cases, it has led directly to protracted wars. In modern times, the attainment of the objectives of the state has been given clear precedence. More recently still, it has become closely tied up with the conviction that one social system or the other is superior to the rest.

Now, however, the problem of atomic energy has emerged as a new factor relating directly to the fate of mankind as a whole. We have reached a stage, a turning point, where each member of the human race must take careful stock of the bonds linking his own fate and that of the rest of mankind, devise the best possible scheme for protecting himself from the threat of atomic energy, and devote far greater energy than heretofore to putting it into practice. This would surely be, at the same time, an impressive first step towards the realization of a human community that would be more directly relevant to mankind's prosperity and happiness.

As a scientist, I feel a responsibility to consider the problem of atomic energy versus mankind still more seriously than others. As a Japanese, I cannot help feeling this question more personally than others. Yet neither fact in any way conflicts with my considering it as a member of the human race.

2   The Role of the Modern Scientist: The Tenth Pugwash Conference [1962]

London in early September was cool, at night almost chilly. No sooner had I returned from Cambridge than I caught cold. Bedridden in my room in the Russell Hotel—my lodgings, and the site of the Tenth Pugwash Conference—I told myself that I must shake the cold off as soon as possible so that I could, at the very least, attend the first day of the conference.

On that day, two hundred persons from thirty-five different

nations were gathered in the conference hall. The ninety-year old Bertrand Russell put in his appearance, walking with a firmer step than I had expected. His expression had a benignity quite unlike its former asperity and reminiscent of some distinguished prelate in whom wisdom and compassion were equally combined. He must have been highly gratified that the Pugwash movement should have grown so greatly during the seven short years since the Russell-Einstein declaration. There had been only twenty-two participants at the first conference in 1957, which meant a tenfold increase in five years—a very welcome development, always provided it did not mean that the basic spirit pervading the movement was being forgotten. It had been Russell's idea that at this conference all the original signatories of the declaration should be present. I had hoped to get someone else to represent Japan, but in such circumstances I had felt obliged to attend in person. Of the ten signatories apart from Russell, three—Einstein, Joliot, and Bridgman—had already passed away. Born and Muller were too sick to attend.

There were only five of us, thus, seated on the platform. The proceedings began with a long speech by Lord Hailsham, science minister of Britain, the host country. His first point was that there was, as yet, no scientific theory of government, and that to create one would be no easy matter. He followed this up with the observation that government was lagging behind the present advanced state of science and technology.

His speech was followed by a dozen or so messages from the heads of various states, the secretary-general of the United Nations, the president of the Royal Society, and the like. Then came Lord Russell's address of which the parts that made the strongest impression on me were the extracts he read from letters sent by the two men who had been unable to attend.

Born wrote: "At the third conference, when I acted as chairman, I interposed the following remark: 'clever, rational ways of thinking are not enough. The danger of mass slaughter . . . can only be overcome by moral conviction, by a determination to replace national prides and prejudices with human love.' Even if I had been able to attend this time, I could only have repeated what I said then.''

Muller wrote: "By now, one is forced to wonder whether there is not too much public support where once there was too little."

On the latter score, there seems so far little cause for worry in Japan, since there are still plenty of people who shake their heads at the very idea of my attending such a conference. Concerning his call for a moral stand, it is identical with a point that we stressed at the Kyoto Conference of Scientists in May; and the paper submitted to the tenth conference by Japan was, in fact, concerned with precisely this moral question. I was encouraged by the fact that the physicist Heitler, an old acquaintance, dealt with the question of science and morality in the broadest sense.

I had the impression, nevertheless, that far from being concerned with the moral question, the majority of scientists who gathered there were actually convinced that they should not delve any farther than keen rational thought would take them. A considerable number among them seemed to feel it was enough to concentrate on the immediate question of disarmament. To do that, however, is to lay emphasis solely on the technological know-how of scientists from the great powers. There is a need from time to time, as for example at this London conference, to broaden the argument and provide a chance for a large number of representatives from the smaller nations to participate.

The questions dealt with included, besides "The Scientist and World Security" and other questions directly related to disarmament, "The Position of the Scientist in Society," "International Cooperation in Science," "Science and Aid to the Developing Nations," and "Science and Education." While each of these questions is undoubtedly important, many of them are being attended to by various organizations of the United Nations. For a problem to be taken up by the Pugwash Conference should automatically mean that it is in some way different. In that respect, I could not repress a certain disappointment.

Another problem concerned the organization of the Pugwash Conference as such. Although a tenfold increase in membership within five years is undoubtedly a matter for rejoicing, it has brought with it new difficulties. Its original nature as a loose organization based on the individual is gradually being altered by the emergence of a new type of organization in which people in areas all over the globe form their own organizations to participate in the movement. In this sense, the Pugwash Conference has come to a kind of turning point in its history.

For the first three days of the conference I forced myself to attend, but on the last two days I retired to bed in my hotel room. Fortunately I was not, as at the Cambridge conference, the sole representative of Japan. Iwao Ogawa had come from Japan, and Susumu Kamefuchi, then at London University, was also attending. To some extent, Ogawa acted as spokesman on my behalf.

A statement was issued at the end of the conference, but this kind of statement has lost a considerable amount of its point compared with the early days of the Pugwash movement. The last passage in the statement itself read, in fact, as follows:

> We are now at a stage in which general statements of principles are not enough; action is needed. . . . We reassert out conviction that the goal of full disarmament and permanent peace is realistic and urgent. This work is truly to be seen as part of a long struggle for the progress of mankind, and it is one in which scientists have a responsible part to play. We call upon scientists everywhere in the world to join us in this task.

As a scientist, a Japanese, a member of the human race, what

192 • ON PEACE

ought I to do—what can I do? How can I reconcile moral duty with the study of theoretical physics? Reading the copy of the declaration brought me by Ogawa, I reflected on such questions and felt my headache growing steadily worse.

My cold persisted until after my return to Japan.

3   The Concept of Peace in the Nuclear Age [1968]

The Rejection of War

Throughout the long history of man, there has probably been no age in which the desire for peace has taken such a strong and persistent hold on men as our own. The hold has increased in direct proportion to the rapid increase in the destructive powers of war. In particular, the emergence of the atomic bomb during the closing days of World War II and the development of the hydrogen bomb following the end of the war have had a decisive influence on the views of war held by a large number of people. As a result, the view that rejects war entirely has found acceptance on an incomparably wider scale than before. This outlook is probably most pervasive and most firmly rooted among the Japanese, who themselves experienced the horror of the atomic bomb and suffered casualties as a result of a hydrogen bomb test. The way in which Japan has, despite various controversies, preserved intact the "peace constitution" that came into being immediately after the end of the war is undoubtedly an outcome of the Japanese people's deep-rooted aversion to war.

If one were to ask oneself, or other people, just what constituted, or should constitute, pacifism, one would undoubtedly get all kinds of different replies. What is almost self-evident, though, is that the rejection of war is, and must be, a common ingredient in every case. Something that is self-evident should need no explanation; yet it might be as well to sum up the essentials of the question before proceeding further.

Prior to the emergence of nuclear weapons, the outlook that rejected war in general was in fact restricted to a permanent minority. Opposition to war—or, on an even more thoroughgoing level, the advocacy of nonviolence and nonresistance—was regarded by the majority of people as completely unrealistic. There was a wide range of shades in the affirmation of war, from the negative view that, while not desirable, it was in some cases inevitable, to the positive view that one should join wholeheartedly in a just war; but the view that in certain cases war was justified was in an absolute majority. With the appearance of nuclear weapons and the steady increase in their power, it has become more and more clear that talk of the total destruction of mankind, far from being a hyperbole, is a real possibility that cannot be overlooked. The total destructive power of the stockpile of nuclear weapons at present held by the United States and the Soviet Union is said to be enough to wipe out the entire human race many dozens of times over. In such a situation there can no longer be any justification for nuclear warfare. The question today, therefore, is whether not only nuclear war but war in general should be rejected or not. In the world as it is today, with the rapid and continuing increase in weapons in general, even a small-scale war—a localized war making no use of nuclear weapons—can inflict a terrifying amount of destruction. Although chemical and biological weapons may not have the decisive force of nuclear weapons, their capacity for widespread slaughter and their cruelty are likely to go on increasing. There is a constant possibility, moreover, that a conventional war might escalate into a nuclear war.

It seems likely, nevertheless, that many people will remain unconvinced even by reasons as cogent as these. The time-honored concept of the "just" war still persists today. War, of course, is waged by nations or groups of nations. Where one of the nations concerned decides that the war is "just," the decision is made in accordance with the view that has the greatest power of influence in determining and carrying out that nation's policies. I will not discuss here, nor do I consider myself qualified to discuss, the particular processes whereby particular persons, or the particular outlook of a particular class of persons, come to exert the greatest influence in policy making. What I do know, though, is that underlying this outlook, whatever it may be, there is always a set of values that is used to justify the war. Even at present, the term "crusade" is still heard in various contexts. The Crusades were originally expeditions aimed at recapturing the holy city of Jerusalem, and it was Christianity, as an absolute set of values, that justified those expeditions. Yet the Muslims against whom they were directed believed equally in Islam as an absolute value system. For the man who believes in neither religion, it is impossible to judge either side as right. Should the Muslims cite their right to legitimate self-defense against aggression, the Christians would doubtless counter by using ancient history to prove that they were only recapturing what was originally their territory.

Today, the rejection of war implies the rejection of all these various types of war. And it stresses the need to settle international disputes by peaceful means, without resort to war. This is precisely what Article Nine of the Japanese constitution is about, and it is for that reason that Japan renounced the right to make war. World history, it is true, affords examples of wars in which, seen from our present vantage point, it is possible to judge that one side or the other was in the right: the wars for the liberation of colonies are a case in point. Yet in the world as it is today, even this kind of war should be replaced entirely by peaceful negotiations. When it is taken as far as this, the idea of the rejection of war acquires a strong air of idealism; if it is to be made more down to earth, it must be accompanied by intensive cooperation on an international scale aimed at enabling those countries that are trying to achieve independence—those that have legitimate reasons for doing so—to achieve it by means other than war.

A major problem in the world today is the widening of the gap between the large nations and the small nations. Most of the many new nations formed when the former colonies acquired their independence following the war are small nations. On the other hand, some of the nations that were already great powers before the war have become more powerful still. By maintaining nuclear weapons, moreover, the great powers have sought to widen the gap between themselves and the nonnuclear nations. The two greatest powers in particular, the United States and the Soviet Union, have an overwhelming superiority over other nations in both military strength and political influence. It is because both of them judged this situation to be extremely advantageous to themselves that they showed such extraordinary enthusiasm for the signing of a nuclear nonproliferation agreement. In practice, not only was the content of the treaty unsatisfactory to the nonnuclear nations, but even the United States and the Soviet Union themselves were to prove unable to maintain the situation as it was. The reason is that the situation was not truly stable, but depended on a balance maintained in a continuing fierce competition, mainly in nuclear arms, between the two principals. Not merely was this astonishingly wasteful for the two nations concerned, and for the whole of mankind, but it was hardly something that could be continued indefinitely.

There had long been talk of the nuclear deterrent strategy—the theory that so long as a situation was maintained in which neither of the two great powers could destroy the other with its first nuclear attack, so that the first to move would inevitably incur enormous losses as a result of a second, retaliatory, strike. Neither side would be able to make the first move, and nuclear war would not occur. This theory becomes increasingly difficult to maintain as nuclear powers other than the big two increase in number. It served as an important factor encouraging the two great powers to promote the signing of a nonproliferation treaty; but it hardly needs profound consideration to see that the deterrent theory is basically mistaken. For nonnuclear nations and for mankind as a whole, nuclear weapons present an unspeakable peril that must be removed completely from the face of the earth. I doubt whether even the two great powers believe that they can maintain the deterrent theory indefinitely. And, as we shall see in the next section, a completely different trend, in which the two great powers themselves are participating, does in fact exist.

Peaceful Coexistence and Complete Overall Disarmament

If one takes the rejection of war as a starting point for today's pacifism, then a minimum requirement for making this into something effective and realistic is the idea of peaceful coexistence. If wars based on the "crusader" outlook discussed in the last section are to be avoided, the first necessity is the idea of peaceful coexistence, the idea that, even where there are basic points of conflict between a set of values espoused by one nation or group of nations and a set of values espoused by another nation or group of nations, neither side will resort to war in the attempt to impose its own values on the other. In today's world, every nation must strive to ensure that nations with mutually contradictory sets of values can go on coexisting in peace. The idea of peaceful coexistence in this sense has been gaining rapidly in strength since the fifties, and seems by now to have become firmly established among a large number of nations and individuals.

If this is the first stop on the route to a truly peaceful world, the next is surely the scrapping of the various nations' armaments. So long as sovereign nations continue to possess arms they will continue their efforts to see that these do not become ineffective. They will reinforce their armaments just as far as their national resources allow. As a result, the arms race will continue indefinitely. Even the idea of the nuclear deterrent is useless in stopping the nuclear arms race. Yet despite this—or rather, precisely because of it—efforts towards disarmament have continued steadily.

At the 1959 General Assembly of the United Nations, an eighty-two nation resolution on "Complete and Overall Disarmament" was presented by all member nations, and unanimously adopted. The resolution was an expression of the determination to work with the main goal of completely doing away with all nations' armaments, conventional arms as well as nuclear weapons. It would be possible of course to sneer at it as the empty phrase-making of politicians and diplomats. Yet whatever motives may have inspired individual representatives to the United Nations, the resolution was, nevertheless, a reflection of a definite human aspiration. Two years later, in 1961, a joint declaration by the United States and the Soviet Union was issued in the form of a report to the United Nations General Assembly. The declaration was a summary of the agreement reached by the two nations, based on the General Assembly resolution, concerning the principles to be observed in carrying out disarmament negotiations involving many nations.

The United Nations Disarmament Committee, which had already been in existence for some time, seems to have been encouraged by the resolution and declaration into stepping up its activities. Two years later again, in 1963, the two nations signed a treaty providing for a partial ban on nuclear tests. Views of the treaty's worth are far from unanimous, but I, at least, welcomed it as a possible first step towards world peace.

Subsequent developments in the world situation, however, betrayed our hopes. A number of conspicuous developments—

the wars in Vietnam and the Middle East, and the invasion of Czechoslovakia—have gone directly counter to the trend towards world peace. The number of nuclear powers, including those with the hydrogen bomb, has increased to five with the addition of China and France. The nuclear arms systems of the United States and Soviet Union have grown steadily more diversified and more enormous. To the already existing atomic submarines carrying nuclear missiles have been added, recently, antiballistic missiles and orbiting nuclear weapons, the appearance of which is serving to make increasingly clear the contradiction implicit in the nuclear deterrent theory, while on the other hand rendering impotent the United Nations Committee for Disarmament. The phrase "a world under a nuclear umbrella" that has come into use lately has an almost eschatological ring. The term "nuclear allergy" has also come into common use to signify what is considered the excessive sensitivity of the Japanese towards nuclear weapons. In fact, though, a nuclear explosion is equally dangerous to both the Japanese and the people of other nations. It is the Japanese, rather, who are normal in showing fear of something frightening; for no amount of insensitivity to nuclear weapons will lessen their danger. Man is in a "nuclear age," in the sense that nuclear weapons exert a dominating influence over human affairs, but we do not, of course, reject the peaceful uses of nuclear energy; we look forward, rather, to a postnuclear world in which nuclear energy, its fangs drawn, will work in the overall service of mankind. At such a time, however, the question of preventing the dangers involved in the radiation and radioactivity that accompany nuclear fission will be increasingly serious. The greatest caution will be required on this score. Thus, there is no ground whatsoever to use such a term as "nuclear allergy." A great deal of attention has been paid recently to environmental pollution, and rightly so. In this case, too, it was not the "allergy," but rather the lack of concern shown so far, indeed, that has contributed to the rapid increase in pollution. Before the nuclear generation of electricity is begun on a large scale, some sure means of handling the radiation involved must be established.

To get back to the main subject: the ten years since the members of the United Nations all joined in clearly setting forth complete and general disarmament as their common goal have made attainment of that goal seem, if anything, more difficult than ever. Such being the case, it may seem completely pointless to many people to think still further, to a time beyond the attainment of complete overall disarmament. Some of them may be convinced, even, that the nuclear age in the sense already described will continue indefinitely, as man's inescapable destiny. It is not a natural phenomenon, however, that nuclear weapons should have been created and a nuclear age should have continued for as long as twenty years. The present vast arms setup with nuclear weapons at its center has been developed by man himself. That being so, it should not be impossible for man to do away with it also by his own efforts. Nuclear weapons, it should always be remembered, do not exist, like some distant star, as a simple object in nature, but are always closely linked with man's own intellectual activity. And the way international politics is conducted is also a manifestation, in a different form, of that same intellectual activity. Those two intellectual activities, moreover, exert an unceasing influence on each other. Is not the difficulty in achieving complete and overall disarmament a result of divorcing it from the question of how international politics is conducted and treating it as a goal in itself? To put it in other words, it is useless to consider complete overall disarmament as the terminal station in itself; even disarmament, I feel, will never be achieved unless it is considered as one goal among others, as part of a larger vision of a world at peace.

The Concept of World Federation in the Nuclear Age

Disputes between nations must be prevented from escalating into war, and a minimum requirement if this is to be achieved is a place for discussion among the nations. In this respect, the United Nations already exists as an organization where the representatives of the nations can discuss and pass resolutions on their problems. It has, moreover, frequently made great contributions to the peaceful settlement of international disputes. But many other cases have conveyed a bitter sense of the United Nations's impotence, especially where the countries concerned in the dispute were nuclear nations. It is obvious to all moreover, that the absence of some nations from the organization—and in particular the exclusion of the People's Republic of China, which accounts for one fourth of the total world population—is a serious failing.

The United Nations Charter was drawn up a little before the end of World War II. In a sense, thus, the United Nations was out of date from the start as a piece of machinery for the settlement of international disputes in the nuclear age. The thing men long for today is, somehow, to pass beyond the nuclear age and create a new and better age. The vision that men cherish of international politics under that new situation must, inevitably, rise far above the stage of the United Nations.

It was shortly after the appearance of the atomic bomb that I first began to feel a strong interest in the idea of world federation, but the idea as such is far from new. It has repeatedly been observed that so long as human society continues to consist of an assembly of many states each with absolute sovereignty over its own affairs, the possibility of war between states or between groups of states will never disappear. And the same observation is frequently followed by the assertion that lasting peace will not come about until some world law is established as a superior authority that will make national sovereignty relative, restricting it and binding it, and until some machinery of world federation is set up in conformity with that law. The origin of this idea cannot be traced to any one proponent; the truth seems to be that it appeared independently at various times in various areas.

In 1879, Azusa Ono, a young man of eighteen born into what had been until recently the Tosa clan, went to Shanghai, then a virtual colony of the Western powers, where he was painfully struck by the poverty of the Chinese inhabitants. In his *Kyūmin Ron*, written at that time, he called for the holding of a world conference with delegates from all the nations of the world, with the aim of establishing a world law and setting up an administrative organization to govern in accordance with that law.

A few years after this, Emori Ueki expounded a similar idea in detail in his *Mujō Seifu Ron*, while Chomin Nakae also, in his *Sansuijin Keirin Mondō*, put the idea of a world federation into the mouth of a student of Western learning. The world federation movement that has become active in Japan since the end of World War II, under the leadership of men such as Yukio Ozaki, Toyohiko Kagawa, and Yasaburo Shimonaka, is at the same time part of a wider international movement and a development of the ideas of world federation that have existed in Japan since the time of Azusa Ono.

As I have already said in the preceding section, despite all the talk of complete and overall disarmament, in practice it proved elusive. An important reason here is the difficulty in solving the question of arms inspection. No country is likely to rush into disarmament without assurances that not only itself but other countries too will carry out disarmament in accordance with their promises. This means that the international machinery for inspection must be given authority superior to that of the sovereignty of individual nations. For this to be recognized the United Nations must cease to be a mere gathering of nations each with absolute authority over its own affairs and acquire the nature of a true world federation. Seen in this light, it becomes clear that the process whereby each nation reduces its armaments and the process whereby world federation is realized should not be separate but should take place—at least in part—concurrently.

If this is so, then the question that arises is how to create an opportunity for the United Nations to shift to that higher plane. One possible method would be for one nation, or a number of nations, to surrender voluntarily a part of its, or their, sovereignty. Japan, though, has already in fact done so in its "peace constitution." As is common knowledge, under Article Nine of that constitution she has unilaterally renounced war as an exercise of her national sovereignty. At the World Congress of the World Conference for World Federation held in Japan in 1963, we appealed to participants from other countries to work to achieve similar declarations by their own countries. The appeal met with much sympathy, and the congress subsequently passed a resolution to this effect.

Article Nine of the constitution, incidentally, bears a close and inseparable connection with the preamble of the same constitution. It is stated in the preamble that "We have determined to preserve our security and existence, trusting in the justice and faith of the peace-loving peoples of the world." It goes on to state: "We believe that no nation is responsible to itself alone, but that laws of political morality are universal; and that obedience to such laws is incumbent upon all nations who would sustain their own sovereignty and justify their sovereign relationship with other nations." It was necessary, if this conviction was not to prove divorced from reality, that human society as a collection of nations possessing arms and the right to belligerency should gradually change into human society as a collection of nations that had cast aside their arms and, renouncing war, were faithfully observing international law. Implicit in it, in other words, one may see the demand that world federation should be realized and the expectation that it would be realized in fact.

There is nothing particularly new about what I have written, and it doubtless contains much that has already been expressed by others. It is my hope, rather, that as many people as possible will feel that this is so. For I believe it to be desirable that the better age to follow the nuclear age should be brought about not by the efforts of a select few but by the workings of the conscience and the good sense that lie, one hopes, in the hearts of nearly all men.

4   Wisdom for Modern Man [1956]

Progress in science has gradually permitted us to understand rationally the world outside ourselves. This so-called outside world in fact embraces everything, including our own persons. Yet at the same time, the progress of science only makes us more clearly aware that there remain in this outside world things that are still unknown to man, things that still lie beyond his rational understanding. In that sense, man lives in an "open" world. Nevertheless, people have tended to underestimate or to forget the fact that the world inside man—the inner world, what is referred to as the world of the mind—is also still, in the same sense, an open world. When this is taken into account, the new significance of the Freudian way of thinking will, I think, be apparent.

It is obvious that unless something rises into the consciousness it cannot become an object of rational thought, of rational consideration. At the same time, everyone will surely agree on the importance of an awareness that living in a world open towards both the exterior and the interior is a special characteristic of man's lot. Yet so long as the Freudian approach goes no further than analysis of the human psychology and neglects the fact that it also has a close bearing on the whole world in which we live—the whole world, including the physical world—it cannot help but remain an inadequate, one-sided way of seeing things.

Questions such as these are really of course the province of specialists in psychology and not of a physicist such as myself. One thing, though, that I would like to emphasize is that when science becomes associated with technology and technology becomes associated with all kinds of interests in the world of man, it is impossible to confine the question to the rational aspect of human nature alone. So long as the nuclear physicist was studying the structure of matter, it did not matter particularly what were his aspirations as a human being or what kind of life he led. One could legitimately consider his ideas and life outside his studies as something quite separate from his research into the structure of matter.

However, from the moment that research into the atom advanced to the point where the use of atomic energy became a practical possibility, one could no longer separate his way of life and thinking as a scientist from his way of life and thinking in other spheres. It became quite impossible to claim that, whatever the purposes to which atomic energy might be put, it was no concern of the research worker himself. The atomic physicist is, of course, only the most striking example; in a very great number of other cases as well—though admittedly to a lesser degree—the point at which the fruits of scientific research begin to have practical applicability marks the intrusion of ethical and moral considerations also.

The development of science, far from encouraging the formation and development of an intergrated view of man in his many different aspects, has tended rather to split and destroy that view.

The differentiation of science into many seperate fields has in specific areas created large numbers of specialists in possession of detailed and reliable knowledge and techniques. As the types of machine made by man have increased in number, and as they have become more complex, the men who use them have gradually been obliged to content themselves with using them simply for their convenience, without bothering to go into how they work. Man, while telling himself that it is he who is using the machine, is by imperceptible stages changing into a creature who cannot live without its aid. We should of course be grateful to the machine for taking over, expanding, and stepping up all kinds of processes hitherto performed by men, but when it begins gradually to usurp even the functions of the human brain one cannot but wonder where it will all end.

The danger that all these trends will combine to destroy the unity of man's various aspects and lead eventually to the loss even of his humanity cannot be denied. It is equally undeniable, I believe, that such a split would deprive man of his happiness. There is much discussion these days of whether science really contributes to human happiness or not, and it is very difficult to supply any confident answer. Whatever may have been true for the scientist of the nineteenth century, for us who happen to be alive in the middle of the twentieth century the flat assertion that science enhances human happiness is difficult to make.

There never was, in fact, any guarantee that science would make mankind happy. Science is a manifestation of man's efforts to open up the unknown world that lies before him, to discover new possibilities. Yet who can know what the world of the unknown holds in store? There is, from the outset, no guarantee that the discovery of new possibilities will make man happy. A discovery may lead to happiness and prosperity, or it may, equally well, lead to a loss of humanity and the destruction of mankind.

It is extremely difficult, of course, to define human happiness. One doubts whether any branch of learning could give the answer; quite probably, in fact, human happiness will remain eternally elusive as an objective of scholarly research. The human emotions have their origins deep in the human heart. In most cases, their beginnings lie in regions beyond the reach of human consciousness and moral reflection. As I have already said, man has things in him that are unknown even to himself. He still retains within him many things that he possessed before man became man. He has them whether he realizes it or not. The human emotions are closely related to these. Thus it is difficult, in discussing human happiness, to make simple scientific assertions that do not take them into account. Every individual has in him things that cannot be simply defined. Thus human emotions, and human happiness in turn, are closely bound up with these indefinables.

In the world of men, it is the things that everyone considers to be right and desirable in principle that fail to be realized, while the things to which there are many objections get done—a fact that is eloquent proof that man is not moved solely by the conscious self, by the self as the object of rational thinking and reflection. Yet this does not mean, either, that we human beings should make light of the reason, the power of rational thought, that we possess. On the contrary, we have the possibility, by letting all the things that lie hidden deep at the bottom of the consciousness rise to the surface, or by sending our reason delving farther into the depths, of gradually bringing a wider area of our humanity within the scope of rational consideration. To save ourselves, by such efforts, from the danger of a loss of, or a split in, our humanity in the world of the future—this, I believe, is the course that man must take from now on. This, it is my personal conviction, is true wisdom for modern man.